AQA GCSE

MATHEMATICS
for Middle sets
TEACHER GUIDE

for **Modular** and **Linear** specifications

Series editor: **Glyn Payne**

Authors:
Gwenllian Burns
Greg Byrd
Lynn Byrd
Crawford Craig
Janet Crawshaw
Fiona Mapp
Avnee Morjaria
Catherine Murphy
Catherine Pate
Glyn Payne
Ian Robinson
Harry Smith

www.pearsonschools.co.uk

✓ Free online support
✓ Useful weblinks
✓ 24 hour online ordering

0845 630 22 22

Longman
Part of Pearson

Longman is an imprint of Pearson Education Limited, a company incorporated in England and Wales, having its registered office at Edinburgh Gate, Harlow, Essex, CM20 2JE. Registered company number: 872828

www.pearsonschoolsandfecolleges.co.uk

Longman is a registered trademark of Pearson Education Limited

Text © Pearson Education Limited 2010

First published 2010

14 13 12 11 10
10 9 8 7 6 5 4 3 2 1

British Library Cataloguing in Publication Data
A catalogue record for this book is available from the British Library

ISBN 978 1 408 23283 5

Copyright notice

All rights reserved. The material in this publication is copyright. Pupil sheets may be freely photocopied for classroom use in the purchasing institution. However, this material is copyright and under no circumstances may copies be offered for sale. If you wish to use the material in any way other than that specified you must apply in writing to the publishers.

Edited by Alex Sharpe and Project One Publishing Solutions, Scotland

Schemes of Work written by Gwenllian Burns

Additional material written by Maggie Rumble

Designed by Pearson Education Ltd

Typeset by K & S Design and Standard Eight Limited www.std8.com

Original illustrations © Pearson Education Ltd 2010

Illustrated by K & S Design

Cover design by Wooden Ark

Cover photo © iStockPhoto/Stuart Berman/

Printed in the UK by Ashford Colour Press Ltd

Acknowledgements

Every effort has been made to contact copyright holders of material reproduced in this book. Any omissions will be rectified in subsequent printings if notice is given to the publishers.

Websites

The websites used in this book were correct and up-to-date at the time of publication. It is essential for teachers to preview each website before using it in class so as to ensure that the URL is still accurate, relevant and appropriate. We suggest that teachers bookmark useful websites and consider enabling students to access them through the school/college intranet.

Contents

An alternative Modular Two-year Scheme of Work (starting with Unit 1) is on the CD-ROM at the back of this Teacher Guide.

Lesson Plans

Unit 1:

Unit 2:

Unit 3:

On the CD-ROM at the back of this Teacher Guide (in PDF and Word format):

- All **Guided Practice Worksheets** referenced to Lesson Plans, with Answer Sheets.
- All the content printed in this Teacher Guide.
- Two-year Modular Scheme of Work starting with Unit 1: overview and full detail.

About this Teacher Guide

All set to make the grade!

AQA GCSE Mathematics for Middle sets Student Book and Practice Book are written by senior examiners and practising teachers to help your students get their best grade in the exams.

This **Teacher Guide for Middle sets** supports your teaching with:

- Simple and straightforward **Modular and Linear Two-year Schemes of Work** – in print and in editable Word files on the CD-ROM in the back of this Teacher Guide

- **Personal Progress Chart** – printed photocopy master – to help each student monitor their own achievement through the course

- **Lesson Plans** for each section of each chapter in the Middle sets Student Book – in print and in editable Word files on the CD-ROM in the back of this Teacher Guide

- **Guided Practice Worksheets** (fully referenced in the Lesson Plans) for students who need a bit more support in class or with homework – as PDFs and in editable Word files on the CD-ROM in the back of this Teacher Guide.

Answers are provided.

See further examples on pp vi–ix

Numbering matches Lesson Plan

Help given to get students started

Worksheets in write-on format

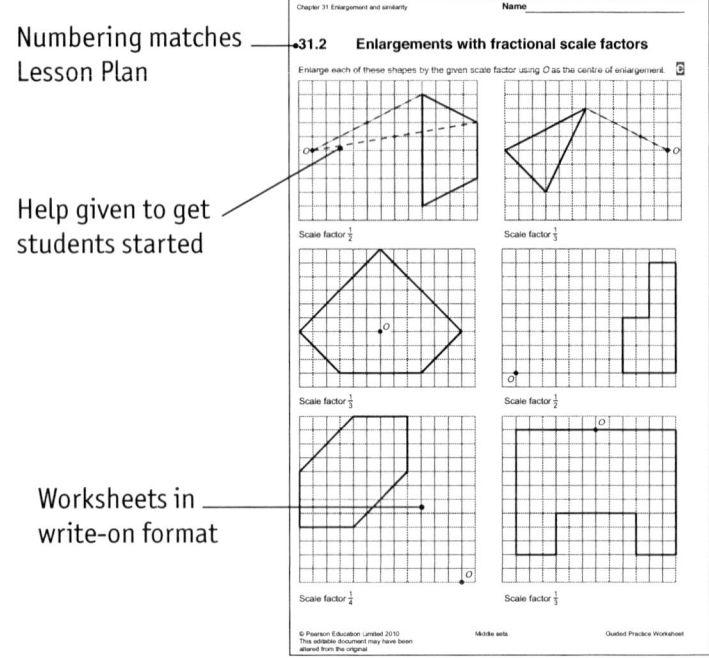

Personal Progress Chart

The grades in AQA GCSE Mathematics for Middle sets show you the level of difficulty for each question (grade E is the easiest and grade B is the hardest).

The Personal Progress Chart helps you track your progress through each Unit of your GCSE course.

How to use the Personal Progress Chart

1. Together with your teacher, agree your target grade for the Unit. Write this in the 'My target grade' space for every Unit chart.

2. At the end of a section, shade in the box that matches the highest grade of question you achieved.

3. As you work through the chapters, you'll see how you're doing compared with your target grade.

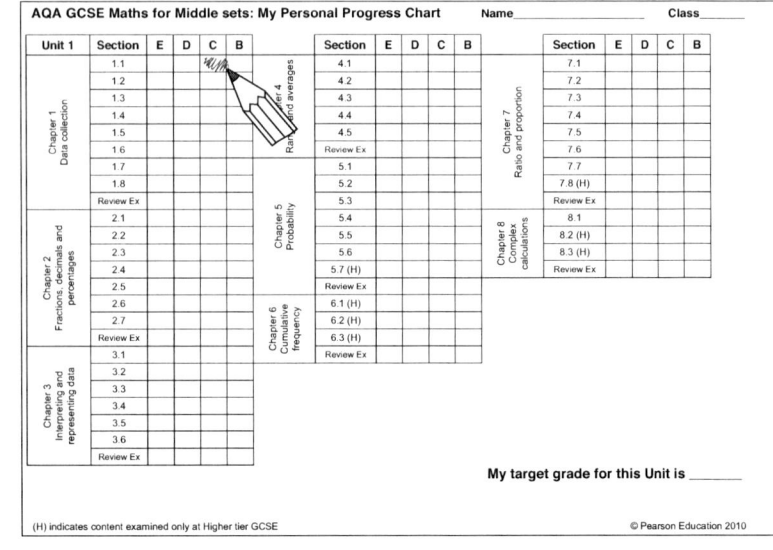

One-year and Three-year Schemes of Work

These are provided as free downloads, from our website at **www.pearsonschools.co.uk/aqagcsemaths**

BBC Active Video clips

In the ActiveTeach whiteboard software for Middle sets, there are plenty of stimulating video clips to support the course. Each one is referenced in the appropriate lesson plan in this Teacher Guide. Topics include:

- Supercross
- Traffic
- Fairground
- Printers
- 3D graphics

Also in the Middle sets ActiveTeach whiteboard software (ISBN 978 1 408232 80 4):

- **Grade Studio** interactive practice for AO1-, AO2- and AO3-type questions
- **'Watch the examiner' video clips** showing how exam questions are marked, with hints on how to get full marks!
- **Topic Tutor** with step-by-step solutions to a selection of questions with accompanying voiceover
- **Lesson Player** to combine and pre-arrange ActiveTeach digital resources with your own, and then play them in sequence through your lesson
- **Class tracker** to record and monitor progress through each resource
- **Exam Café full** of revision plans, activities and handouts for exam preparation

AQA GCSE Mathematics Assessment Pack (ISBN 978 1 408232 84 2)

Covering both tiers of the whole GCSE course in one pack:

- **Course Entry Tests** to help set your students for GCSE
- **Chapter Tests** for all three Student Books
- **Modular Exam Practice Papers** for Higher tier and Foundation tier, full of questions reflecting the make-up of the new Unit exams
- **Linear Exam Practice Papers** for Higher tier and Foundation tier, full of questions reflecting the make-up of the new Papers
- **Re-sit Practice Papers** for Modular and Linear
- **Class Tracker** to help record students' progress and help decide which tier to enter them in Unit exams.

Keeping in touch...

Subscribe to the blog at http://aqagcsemaths.wordpress.com/ to get news and commentary as the AQA GCSE Maths course and these resources move forward.

ANSWERS

- Student Book answers are printed in the back of the Student Book.
- Practice Book answers are included in the Practice Book Digital Edition and they are provided as free downloads from our website at www.pearsonschools.co.uk/aqagcsemaths
  IN ALL CASES PLEASE USE THIS PASSWORD: **teachersupport**
- Guided Practice Worksheet answers are on the CD-ROM in the back of this Teacher Guide.

SAMPLE Guided Practice Worksheet (Statistics) - plus Answers (shown reduced from A4)
All Guided Practice Worksheets (plus Answers) are available on the CD-ROM in the back of this Teacher Guide.

Answer sheet

3.6 Frequency polygons

1) The frequency table shows the time, t, taken by a sample of students to solve a maths puzzle.

Time taken by a sample of students to solve a maths puzzle

Time, t (minutes)	Frequency	Mid-point
$0 \leq t < 10$	6	5
$10 \leq t < 20$	10	15
$20 \leq t < 30$	12	25
$30 \leq t < 40$	4	**35**
$40 \leq t < 50$	2	**45**

On the grid, draw a frequency polygon of this data. The first point has been plotted for you.

Hint: Plot the frequency against the mid-point of the class interval. Remember to join the points with straight lines.

2) The frequency table shows the weight, w, of some Year 8 students.

Weight of a sample of Year 8 students

Weight, w (kg)	Frequency	Mid-point
$20 \leq w < 30$	6	25
$30 \leq w < 40$	12	35
$40 \leq w < 50$	18	**45**
$50 \leq w < 60$	14	**55**
$60 \leq w < 70$	10	**65**
$70 \leq w < 80$	5	**75**

On the grid, draw a frequency polygon of this data.

3) The table shows the waiting time, t, at a supermarket checkout.

Waiting time at a supermarket checkout

Time, t (minutes)	Frequency	Mid-point
$0 \leq t < 2$	10	1
$2 \leq t < 4$	12	3
$4 \leq t < 6$	8	5
$6 \leq t < 8$	8	7
$8 \leq t < 10$	4	9
$10 \leq t < 12$	1	11

On the grid, draw a frequency polygon of this data.

Middle sets   Guided Practice Worksheet Answers

Name _____

3.6 Frequency polygons

1) The frequency table shows the time, t, taken by a sample of students to solve a maths puzzle.

Time taken by a sample of students to solve a maths puzzle

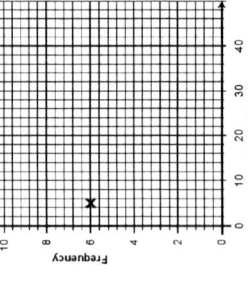

Time, t (minutes)	Frequency	Mid-point
$0 \leq t < 10$	6	5
$10 \leq t < 20$	10	15
$20 \leq t < 30$	12	25
$30 \leq t < 40$	4	
$40 \leq t < 50$	2	

On the grid, draw a frequency polygon of this data. The first point has been plotted for you.

Hint: Plot the frequency against the mid-point of the class interval. Remember to join the points with straight lines.

2) The frequency table shows the weight, w, of some Year 8 students.

Weight of a sample of Year 8 students

Weight, w (kg)	Frequency	Mid-point
$20 \leq w < 30$	6	25
$30 \leq w < 40$	12	35
$40 \leq w < 50$	18	
$50 \leq w < 60$	14	
$60 \leq w < 70$	10	
$70 \leq w < 80$	5	

On the grid, draw a frequency polygon of this data.

3) The table shows the waiting time, t, at a supermarket checkout.

Waiting time at a supermarket checkout

Time, t (minutes)	Frequency	Mid-point
$0 \leq t < 2$	10	
$2 \leq t < 4$	12	
$4 \leq t < 6$	8	
$6 \leq t < 8$	8	
$8 \leq t < 10$	4	
$10 \leq t < 12$	1	

On the grid, draw a frequency polygon of this data.

Middle sets   Guided Practice Worksheet

VI

SAMPLE Guided Practice Worksheet (Number) - plus Answers (shown reduced from A4)
All Guided Practice Worksheets (plus Answers) are available on the CD-ROM in the back of this Teacher Guide.

Name _____

16.4   Repeated percentage change

You **must** show your working in all your answers.

1) Alex invests £2000 in a savings account at 10% per annum compound interest.
    How much will he have in the account after 2 years?

Year 1:   Write down the amount at start of year 1.   _____

    Work out the interest in year 1.   _____

Year 2:   Add the interest in year 1 to the amount at the start
    of year 1 to find the amount at start of year 2.   _____

    Work out the interest in year 2.   _____

    Add the interest in year 2 to the amount at the start
    of year 2 to find the amount in account at the end of year 2.   _____

2) Helen invests £750 in a savings account with 10% p.a.
    How much will she have in the account after 1 year?

    _____

3) Siamak invests £400 in a savings account at 10% p.a. compound interest.
    How much will he be in the account after 2 years?

    _____

4) The number of rats in a sewer increases at a rate of 20% each month.
    There were 50 rats 2 months ago. How many are there now?

    _____

5) The population of a city is 250 000 and it is decreasing at the rate of 10% p.a.
    What will the population be in 2 years' time?

    Hint: The population is decreasing, so subtract 10%.

6) The value of a car depreciates by 20% each year. It was worth £8000 two years ago.
    How much is it worth now?

    Hint: 'Depreciates' means 'gets less'.

    _____

16.4   Repeated percentage change

You must show your working in all your answers.

1) Alex invests £2000 in a savings account at 10% per annum compound interest.
    How much will he have in the account after 2 years?

Year 1:   Write down the amount at start of year 1.   £2000

    Work out the interest in year 1.   10% of £2000 = £200

Year 2:   Add the interest in year 1 to the amount at the start
    of year 1 to find the amount at start of year 2.   £2000 + £200 = £2200

    Work out the interest in year 2.   10% of £2200 = £220

    Add the interest in year 2 to the amount at the start
    of year 2 to find the amount in account at the end of year 2.   £2200 + £220 = £2420

2) Helen invests £750 in a savings account with 10% p.a.
    How much will she have in the account after 1 year?

    10% of £750 = £75
    So after 1 year the amount in the account will be £750 + £75 = £825.

3) Siamak invests £400 in a savings account at 10% p.a. compound interest.
    How much will he be in the account after 2 years?

    After 1 year the amount in the account will be £400 + (10% of £400) = £440.
    After 2 years the amount in the account will be £440 + (10% of £440) = £484.

4) The number of rats in a sewer increases at a rate of 20% each month.
    There were 50 rats 2 months ago. How many are there now?

    After 1 month the number of rats was 50 + (20% of 50) = 60.
    After 2 months the number of rats is 60 + (20% of 60) = 72.

5) The population of a city is 250 000 and it is decreasing at the rate of 10% p.a.
    What will the population be in 2 years' time?

    Hint: The population is decreasing, so subtract 10%.

    After 1 year the population will be 250 000 − (10% of 250 000) = 225 000.
    After 2 years the population will be 225 000 − (10% of 225 000) = 202 500.

6) The value of a car depreciates by 20% each year. It was worth £8000 two years ago.
    How much is it worth now?

    Hint: 'Depreciates' means 'gets less'.

    After 1 year the car was worth £8000 − (20% of £8000) = £6400.
    After 2 years the car is now worth £6400 − (20% of £6400) = £5120.

SAMPLE Guided Practice Worksheet (Algebra) - plus Answers (shown reduced from A4)
All Guided Practice Worksheets (plus Answers) are available on the CD-ROM in the back of this Teacher Guide.

18.2 Plotting straight-line graphs

a) i) Complete the tables of values for these four straight lines.
    ii) Plot each line on the grid.

$y = 2x + 2$

x	-2	-1	0	1	2	3
y	-2				4	

$y = 2x - 3$

x	-2	-1	0	1	2	3
y		-5			1	

$y = 3x + 2$

x	-2	-1	0	1	2	3
y		-1		2		

$y = 4 - 2x$

x	-2	-1	0	1	2	3
y			4			-2

b) Which line crosses the y-axis at the highest point?

c) Which lines cross the y-axis at the same point? _____

d) Which lines passes through the point (6, 9)? _____

18.2 Plotting straight-line graphs

a) i) Complete the tables of values for these four straight lines.
    ii) Plot each line on the grid.

$y = 2x + 2$

x	-2	-1	0	1	2	3
y	-2	0	2	4	6	8

$y = 2x - 3$

x	-2	-1	0	1	2	3
y	-7	-5	-3	-1	1	3

$y = 3x + 2$

x	-2	-1	0	1	2	3
y	-4	-1	2	5	8	11

$y = 4 - 2x$

x	-2	-1	0	1	2	3
y	8	6	4	2	0	-2

b) Which line crosses the y-axis at the highest point?
$y = 4 - 2x$

c) Which lines cross the y-axis at the same point?
$y = 3x + 2$ and $y = 2x + 2$

d) Which lines passes through the point (6, 9)?
$y = 2x - 3$

SAMPLE Guided Practice Worksheet (Geometry) - plus Answers (shown reduced from A4)

All Guided Practice Worksheets (plus Answers) are available on the CD-ROM in the back of this Teacher Guide.

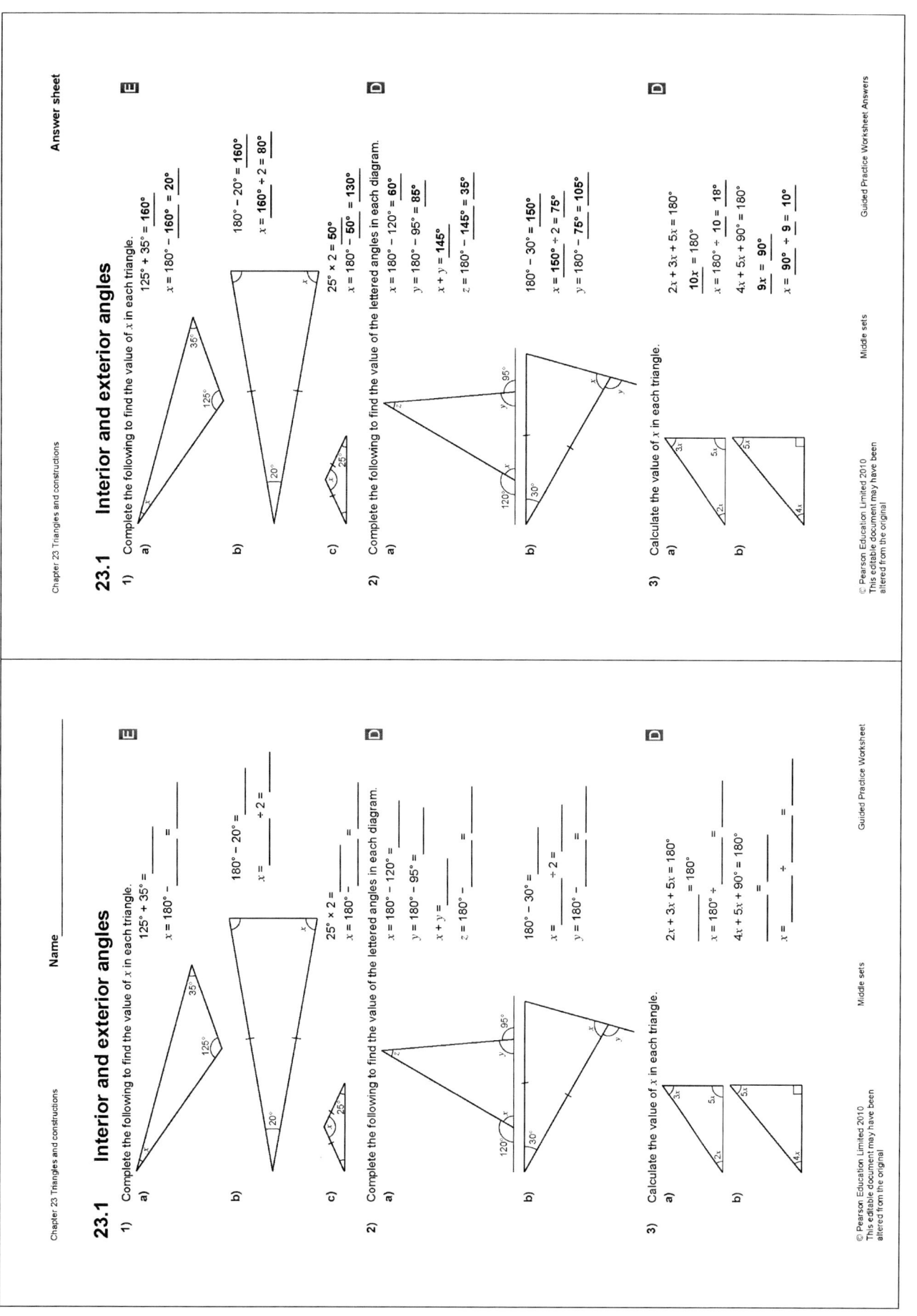

Answer sheet

23.1   Interior and exterior angles

1) Complete the following to find the value of x in each triangle.

a)

$125° + 35° = \underline{\textbf{160°}}$

$x = 180° - \underline{\textbf{160°}} = \underline{\textbf{20°}}$

b)

$180° - 20° = \underline{\textbf{160°}}$

$x = \underline{\textbf{160°}} \div 2 = \underline{\textbf{80°}}$

c)

$25° \times 2 = \underline{\textbf{50°}}$

$x = 180° - \underline{\textbf{50°}} = \underline{\textbf{130°}}$

2) Complete the following to find the value of the lettered angles in each diagram.

a)

$x = 180° - 120° = \underline{\textbf{60°}}$

$y = 180° - 95° = \underline{\textbf{85°}}$

$x + y = \underline{\textbf{145°}}$

$z = 180° - 145° = \underline{\textbf{35°}}$

b)

$180° - 30° = \underline{\textbf{150°}}$

$x = \underline{\textbf{150°}} \div 2 = \underline{\textbf{75°}}$

$y = 180° - 75° = \underline{\textbf{105°}}$

3) Calculate the value of x in each triangle.

a)

$2x + 3x + 5x = 180°$

$\underline{\textbf{10}}x = 180°$

$x = 180° \div \underline{\textbf{10}} = \underline{\textbf{18°}}$

b)

$4x + 5x + 90° = 180°$

$\underline{\textbf{9}}x = \underline{\textbf{90°}}$

$x = \underline{\textbf{90°}} \div \underline{\textbf{9}} = \underline{\textbf{10°}}$

© Pearson Education Limited 2010
This editable document may have been altered from the original

Middle sets   Guided Practice Worksheet Answers

Name _____

23.1   Interior and exterior angles

1) Complete the following to find the value of x in each triangle.

a)

$125° + 35° = $ _____

$x = 180° - $ _____ $= $ _____

b)

$180° - 20° = $ _____

$x = $ _____ $\div 2 = $ _____

c)

$25° \times 2 = $ _____

$x = 180° - $ _____ $= $ _____

2) Complete the following to find the value of the lettered angles in each diagram.

a)

$x = 180° - 120° = $ _____

$y = 180° - 95° = $ _____

$x + y = $ _____

$z = 180° - $ _____ $= $ _____

b)

$180° - 30° = $ _____

$x = $ _____ $\div 2 = $ _____

$y = 180° - $ _____ $= $ _____

3) Calculate the value of x in each triangle.

a)

$2x + 3x + 5x = 180°$

_____ $= 180°$

$x = 180° \div $ _____ $= $ _____

b)

$4x + 5x + 90° = 180°$

_____ $= $ _____

$x = $ _____ $\div $ _____ $= $ _____

© Pearson Education Limited 2010
This editable document may have been altered from the original

Middle sets   Guided Practice Worksheet

AQA GCSE Maths for Middle sets: My Personal Progress Chart

Name _____ Class _____

Unit 1	Section	E	D	C	B
Chapter 1 Data collection	1.1				
	1.2				
	1.3				
	1.4				
	1.5				
	1.6				
	1.7				
	1.8				
	Review Ex				
Chapter 2 Fractions, decimals and percentages	2.1				
	2.2				
	2.3				
	2.4				
	2.5				
	2.6				
	2.7				
	Review Ex				
Chapter 3 Interpreting and representing data	3.1				
	3.2				
	3.3				
	3.4				
	3.5				
	3.6				
	Review Ex				

	Section	E	D	C	B
Chapter 4 Range and averages	4.1				
	4.2				
	4.3				
	4.4				
	4.5				
	Review Ex				
Chapter 5 Probability	5.1				
	5.2				
	5.3				
	5.4				
	5.5				
	5.6				
	5.7 (H)				
	Review Ex				
Chapter 6 Cumulative frequency	6.1 (H)				
	6.2 (H)				
	6.3 (H)				
	Review Ex				

	Section	E	D	C	B
Chapter 7 Ratio and proportion	7.1				
	7.2				
	7.3				
	7.4				
	7.5				
	7.6				
	7.7				
	7.8 (H)				
	Review Ex				
Chapter 8 Complex calculations	8.1				
	8.2 (H)				
	8.3 (H)				
	Review Ex				

My target grade for this Unit is _____

(H) indicates content examined only at Higher tier GCSE

© Pearson Education 2010

AQA GCSE Maths for Middle sets: My Personal Progress Chart

Name _____ Class _____

Unit 2

Chapter	Section	E	D	C	B
Chapter 9 Number skills	9.1				
	9.2				
	9.3				
	Review Ex				
Chapter 10 Factors, powers and standard form	10.1				
	10.2				
	10.3				
	10.4 (H)				
	10.5 (H)				
	10.6 (H)				
	Review Ex				
Chapter 11 Basic rules of algebra	11.1				
	11.2				
	11.3				
	11.4				
	11.5				
	11.6 (H)				
	Review Ex				
Chapter 12 Fractions	12.1				
	12.2				
	12.3				
	12.4				
	12.5				
	12.6				
	12.7				
	Review Ex				

Chapter	Section	E	D	C	B
Chapter 13 Decimals	13.1				
	13.2				
	13.3				
	13.4 (H)				
	Review Ex				
Chapter 14 Equations and inequalities	14.1				
	14.2				
	14.3				
	14.4				
	14.5 (H)				
	14.6 (H)				
	14.7 (H)				
	Review Ex				
Chapter 15 Indices and formulae	15.1				
	15.2				
	15.3				
	15.4				
	15.5				
	15.6 (H)				
	Review Ex				
Chapter 16 Percentages	16.1				
	16.2				
	16.3				
	16.4				
	16.5 (H)				
	Review Ex				

Chapter	Section	E	D	C	B
Chapter 17 Sequences and proof	17.1				
	17.2				
	17.3				
	17.4				
	17.5				
	17.6				
	17.7				
	17.8				
	Review Ex				
Chapter 18 Linear graphs	18.1				
	18.2				
	18.3				
	18.4				
	18.5				
	18.6 (H)				
	18.7 (H)				
	Review Ex				
Chapter 19 Quadratic equations	19.1 (H)				
	19.2 (H)				
	19.3 (H)				
	19.4 (H)				
	Review Ex				

My target grade for this Unit is _____

© Pearson Education 2010

(H) indicates content examined only at Higher tier GCSE

AQA GCSE Maths for Middle sets: My Personal Progress Chart

Name _____   Class _____

Unit 3

Chapter	Section	E	D	C	B
Chapter 20 Number skills revisited					
Chapter 21 Angles	21.1				
	21.2				
	21.3				
	Review Ex				
Chapter 22 Measurement 1	22.1				
	22.2				
	22.3				
	22.4				
	Review Ex				
Chapter 23 Triangles and constructions	23.1				
	23.2				
	23.3				
	Review Ex				
Chapter 24 Equations, formulae and proof	24.1				
	24.2				
	24.3				
	Review Ex				
Chapter 25 Quadrilaterals and other polygons	25.1				
	25.2				
	25.3				
	25.4				
	Review Ex				

Chapter	Section	E	D	C	B
Chapter 26 Perimeter, area and volume	26.1				
	26.2				
	26.3				
	Review Ex				
Ch 27 3-D objects	27.1				
	Review Ex				
Chapter 28 Reflection, translation and rotation	28.1				
	28.2				
	28.3				
	28.4 (H)				
	Review Ex				
Chapter 29 Circles and cylinders	29.1				
	29.2				
	29.3				
	Review Ex				
Chapter 30 Measurement 2	30.1				
	30.2				
	30.3 (H)				
	30.4 (H)				
	Review Ex				
Chapter 31 Enlargement and similarity	31.1				
	31.2 (H)				
	31.3 (H)				
	Review Ex				

Chapter	Section	E	D	C	B
Chapter 32 Non-linear graphs	32.1				
	32.2				
	32.3 (H)				
	Review Ex				
Chapter 33 Constructions and loci	33.1				
	33.2				
	Review Ex				
Chapter 34 Pythagoras' theorem	34.1				
	34.2				
	34.3				
	34.4				
	Review Ex				
Chapter 35 Trigonometry	35.1 (H)				
	35.2 (H)				
	35.3 (H)				
	35.4 (H)				
	35.5 (H)				
	Review Ex				
Chapter 36 Circle theorems	36.1 (H)				
	36.2 (H)				
	Review Ex				

My target grade for this Unit is _____

(H) indicates content examined only at Higher tier GCSE

© Pearson Education 2010

Longman AQA GCSE Maths Two-year Modular Scheme of Work Starting with Unit 2 – OVERVIEW for Middle sets taking Higher Tier

Notes to Teacher:

1. Because Unit 2 has 60% AO1 questions, compared with 40% AO1 for Unit 1, we think that schools might start teaching U2 rather than U1, to give a gentler start to the course. Also, because taking both U2 and U1 exams in Y10 is a time-challenge, GCSE teaching might need to begin in the summer term of Y9.

Please note that our Scheme of Work starting the course with Unit 1 is provided on the CD-ROM in the back of this Teacher Guide.

2. At a minimum, aim for completion of Chapters 9–11 by the end of Year 9.

3. At a minimum, aim for completion of Chapters 12–16 by the October half-term of Year 10.

4. The assumption is that the March and June exams are at the beginning of March and June respectively.

5. Unit 1 revision will need to include the earlier work on ratio (Chapter 7).

6. ***Bold italic*** type shows content that will only be examined at Higher Tier GCSE.

7. Middle sets students are often entered at Higher for Unit 2 and Unit 1, and at Foundation for Unit 3. A 'good Grade C' should be achievable by this route.

	Chapter	Teaching hours	Grades	AQA Modular specification reference
				UNIT 2: Number and Algebra
Y9 SUMMER TERM	9. Number skills	2 [H]	E, D, C	Working with numbers and the number system: N1.2, N1.4, N1.4h
	10. Factors, powers and standard form	6 [H]	E, D, C, B	Working with numbers and the number system: N1.6, N1.7, N1.8, N1.9, N1.9h, N1.10h
	11. Basic rules of algebra	4 [H]	E, D, C, B	The Language of Algebra: N4.1 Expressions and Equations: N5.1, N5.1h
Y10 AUTUMN TERM	12. Fractions	6 [H]	E, D, C, B	Working with numbers and the number system: N1.2, N1.3, N1.5 Fractions, Decimals and Percentages: N2.1, N2.2, N2.7
	13. Decimals	3 [H]	E, D, C, B	Working with numbers and the number system: N1.1, N1.2 Fractions, Decimals and Percentages: N2.3, N2.4
	14. Equations and inequalities	7 [H]	E, D, C, B	Expressions and Equations: N5.4, N5.4h, N5.7, N5.7h
	15. Indices and formulae	6 [H]	E, D, C, B	Working with numbers and the number system: N1.9 The Language of Algebra: N4.2 Expressions and Equations: N5.6
	16. Percentages	5 [H]	E, D, C, B	Fractions, Decimals and Percentages: N2.5, N2.7, N2.7h
	17. Sequences and proof	6 [H]	E, D, C	Expressions and Equations: N5.9 Sequences, Functions and Graphs: N6.1, N6.2
	7. Ratio and proportion	5 [H]	E, D, C, B	Ratio and Proportion: N3.1, N3.2, N3.3
	18. Linear graphs	7 [H]	E, D, C, B	Expressions and Equations: N5.4h, N5.7h Sequences, Functions and Graphs: N6.3, N6.4, N6.5h, N6.6h, N6.11, N6.12
	19. Quadratic equations	6 [H]	B	Expressions and Equations: N5.2h, N5.5h
				UNIT 1: Statistics and Number
Y10 SPRING TERM	1. Data collection	5 [H]	E, D, C	The Data Handling Cycle: S1 Data Collection: S2.1, S2.2, S2.3, S2.4, S2.5 Data presentation and analysis: S3.1

1

© Pearson Education Limited 2010

Term	Topic	Hours	Grades	Specification references
Y10 SPRING TERM	2. Fractions, decimals and percentages	4 [H]	E, D, C	Working with numbers and the number system: N1.14; Fractions, Decimals and Percentages: N2.6, N2.7
	UNIT 2 REVISION FOR MARCH EXAM (8 HOURS)			
	3. Interpreting and representing data	6 [H]	E, D, C	Data presentation and analysis: S3.2; Data Interpretation: S4.2, S4.3
	4. Range and averages	4 [H]	E, D, C	Data presentation and analysis: S3.3; Data Interpretation: S4.1
	5. Probability	7 [H]	E, D, C, B	Data presentation and analysis: S3.1; Probability: S5.1, S5.2, S5.3, S5.4, S5.5h, S5.6h, S5.7, S5.8, S5.9
Y10 SUMMER TERM	*6. Cumulative frequency*	6 [H]	B	Data presentation and analysis: S3.2h, S3.3h; Data Interpretation: S4.4
	8. Complex calculations	5 [H]	C, B	Working with numbers and the number system: N1.10h; Fractions, Decimals and Percentages: N2.7h
	UNIT 1 REVISION FOR JUNE EXAM (6 HOURS)			
	UNIT 3: Geometry and Algebra			
Y11 AUTUMN TERM	20. Number skills revisited	2 [H]		Working with numbers and the number system: N1.3, N1.4, N1.14; Fractions, decimals and Percentages: N2.1, N2.5, N2.7; Ratio and Proportion: N3.1
	21. Angles	3 [H]	E, D, C	Properties of angles and shapes: G1.1, G1.2; Measures and Construction: G3.1, G3.6
	22. Measurement 1	3 [H]	E, C	Working with numbers and the number system: N1.3; Measures and Construction: G3.1, G3.3, G3.4
	23. Triangles and constructions	3 [H]	E, D, C	Properties of angles and shapes: G1.1, G1.2, G1.8; Measures and Construction: G3.9, G3.10
	24. Equations, formulae and proof	5 [H]	D, C, B	The Language of Algebra: N4.2; Expressions and Equations: N5.1, N5.4, N5.6, N5.8; Geometrical reasoning and calculation: G2.3, G2.3h
	25. Quadrilaterals and other polygons	4 [H]	E, D, C	Expressions and Equations: N5.4; Sequences, Functions and Graphs: N6.3; Properties of angles and shapes: G1.2, G1.3, G1.4
	U1/U2 REVISION FOR NOVEMBER RE-SITS (5 HOURS)			
	26. Perimeter, area and volume	4 [H]	E, D, C	Mensuration: G4.1, G4.4
	27. 3-D objects	1 [H]	E, D	Geometrical reasoning and calculation: G2.4
	28. Reflection, translation and rotation	5 [H]	E, D, C	Properties of angles and shapes: G1.6, G1.7; Vectors: G5.1

	Topic	Hours	Grade	Content
Y11 AUTUMN TERM	29. Circles and cylinders	6 [H]	D, C	Properties of angles and shapes: G1.5 Geometrical reasoning and calculation: G2.4 Mensuration: G4.1, G4.3, G4.4
	30. Measurement 2	4 [H]	D, C, B	Measures and Construction: G3.4, G3.7
	31. Enlargement and similarity	5 [H]	E, D, C, B	Properties of angles and shapes: G1.7, G1.7h, G1.8 Measures and Construction: G3.2
Y11 SPRING TERM	32. Non-linear graphs	6 [H]	D, C, B	Expressions and Equations: N5.2h, N5.5h Sequences, Functions and Graphs: N6.7h, N6.8h, N6.11h, N6.13
	33. Constructions and loci	4 [H]	C	Measures and Construction: G3.8, G3.10, G3.11
	U1/U2 REVISION FOR MARCH RE-SITS (5 HOURS)			
	34. Pythagoras' theorem	4 [H]	C	Geometrical reasoning and calculation: G2.1
	35. Trigonometry	7 [H]	B	Working with numbers and the number system: N1.14h Geometrical reasoning and calculation: G2.2h Measures and Construction: G3.6
	36. Circle theorems	4 [H]	B	Properties of angles and shapes: G1.5h
Y11 SUMMER TERM				**UNIT 3 REVISION FOR JUNE EXAM (19 HOURS)**

© Pearson Education Limited 2010

Longman AQA GCSE Maths Two-year Modular Scheme of Work Starting with Unit 2 – OVERVIEW for Middle sets taking Foundation Tier

Notes to Teacher:

1. Because Unit 2 has 60% AO1 questions, compared with 40% AO1 for Unit 1, we think that schools might start teaching U2 rather than U1, to give a gentler start to the course. Also, because taking both U2 and U1 exams in Y10 is a time-challenge, GCSE teaching might need to begin in the summer term of Y9.

Please note that our Scheme of Work starting the course with Unit 1 is provided on the CD-ROM in the back of this Teacher Guide.

2. At a minimum, aim for completion of Chapters 9–11 by the end of Year 9.

3. At a minimum, aim for completion of Chapters 12–15 by the October half-term of Year 10.

4. The assumption is that the March and June exams are at the beginning of March and June respectively.

5. Unit 1 revision will need to include the earlier work on ratio (Chapter 7).

6. ***Bold italic*** type shows content that will only be examined at Higher Tier GCSE.

7. Middle sets students are often entered at Higher for Unit 2 and Unit 1, and at Foundation for Unit 3. A 'good Grade C' should be achievable by this route.

Term	Chapter	Teaching hours	Grades	AQA Modular specification reference
				UNIT 2: Number and Algebra
Y9 SUMMER TERM	9. Number skills	4 [F]	E, D, C	Working with numbers and the number system: N1.2, N1.4, N1.4h
	10. Factors, powers and standard form	3 [F]	E, D, C, B	Working with numbers and the number system: N1.6, N1.7, N1.8, N1.9, N1.9h, N1.10h
	11. Basic rules of algebra	6 [F]	E, D, C, B	The Language of Algebra: N4.1 Expressions and Equations: N5.1, N5.1h
Y10 AUTUMN TERM	12. Fractions	8 [F]	E, D, C, B	Working with numbers and the number system: N1.2, N1.3, N1.5 Fractions, Decimals and Percentages: N2.1, N2.2, N2.7
	13. Decimals	3 [F]	E, D, C, B	Working with numbers and the number system: N1.1, N1.2 Fractions, Decimals and Percentages: N2.3, N2.4
	14. Equations and inequalities	7 [F]	E, D, C, B	Expressions and Equations: N5.4, N5.4h, N5.7, N5.7h
	15. Indices and formulae	6 [F]	E, D, C, B	Working with numbers and the number system: N1.9 The Language of Algebra: N4.2 Expressions and Equations: N5.6
	16. Percentages	6 [F]	E, D, C, B	Fractions, Decimals and Percentages: N2.5, N2.7, N2.7h
	17. Sequences and proof	8 [F]	E, D, C	Expressions and Equations: N5.9 Sequences, Functions and Graphs: N6.1, N6.2
	7. Ratio and proportion	6 [F]	E, D, C, B	Ratio and Proportion: N3.1, N3.2, N3.3
	18. Linear graphs	7 [F]	E, D, C, B	Expressions and Equations: N5.4h, N5.7h Sequences, Functions and Graphs: N6.3, N6.4, N6.5h, N6.6h, N6.11, N6.12
Y10 SPRING TERM	*19. Quadratic equations*	0 [F]	B	Expressions and Equations: N5.2h, N5.5h
				UNIT 1: Statistics and Number
	1. Data collection	6 [F]	E, D, C	The Data Handling Cycle: S1 Data Collection: S2.1, S2.2, S2.3, S2.4, S2.5 Data presentation and analysis: S3.1

UNIT 2 REVISION FOR MARCH EXAM (8 HOURS)

Term	Topic	Hours	Grades	Specification references
Y10 SPRING TERM	2. Fractions, decimals and percentages	6 [F]	E, D, C	Working with numbers and the number system: N1.14 / Fractions, Decimals and Percentages: N2.6, N2.7
	3. Interpreting and representing data	8 [F]	E, D, C	Data presentation and analysis: S3.2 / Data Interpretation: S4.2, S4.3
	4. Range and averages	6 [F]	E, D, C	Data presentation and analysis: S3.3 / Data Interpretation: S4.1
	5. Probability	8 [F]	E, D, C, B	Data presentation and analysis: S3.1 / Probability: S5.1, S5.2, S5.3, S5.4, S5.5h, S5.6h, S5.7, S5.8, S5.9
	6. Cumulative frequency	0 [F]	B	Data presentation and analysis: S3.2h, S3.3h / Data Interpretation: S4.4
	8. Complex calculations	2 [F]	C, B	Working with numbers and the number system: N1.10h / Fractions, Decimals and Percentages: N2.7h

UNIT 1 REVISION FOR JUNE EXAM (6 HOURS)

UNIT 3: Geometry and Algebra

Term	Topic	Hours	Grades	Specification references
Y10 SUMMER TERM	20. Number skills revisited	3 [F]		Working with numbers and the number system: N1.3, N1.4, N1.14 / Fractions, decimals and Percentages: N2.1, N2.5, N2.7 / Ratio and Proportion: N3.1
	21. Angles	5 [F]	E, D, C	Properties of angles and shapes: G1.1, G1.2 / Measures and Construction: G3.1, G3.6
	22. Measurement 1	4 [F]	E, C	Working with numbers and the number system: N1.3 / Measures and Construction: G3.1, G3.3, G3.4
Y11 AUTUMN TERM	23. Triangles and constructions	4 [F]	E, D, C	Properties of angles and shapes: G1.1, G1.2, G1.8 / Measures and Construction: G3.9, G3.10

U1/U2 REVISION FOR NOVEMBER RE-SITS (5 HOURS)

Term	Topic	Hours	Grades	Specification references
	24. Equations, formulae and proof	6 [F]	D, C, B	The Language of Algebra: N4.2 / Expressions and Equations: N5.1, N5.4, N5.6, N5.8 / Geometrical reasoning and calculation: G2.3, G2.3h
	25. Quadrilaterals and other polygons	6 [F]	E, D, C	Expressions and Equations: N5.4 / Sequences, Functions and Graphs: N6.3 / Properties of angles and shapes: G1.2, G1.3, G1.4
	26. Perimeter, area and volume	6 [F]	E, D, C	Mensuration: G4.1, G4.4
	27. 3-D objects	2 [F]	E, D	Geometrical reasoning and calculation: G2.4
	28. Reflection, translation and rotation	6 [F]	E, D, C	Properties of angles and shapes: G1.6, G1.7 / Vectors: G5.1
Y11 SPRING TERM	29. Circles and cylinders	7 [F]	D, C	Properties of angles and shapes: G1.5 / Geometrical reasoning and calculation: G2.4 / Mensuration: G4.1, G4.3, G4.4

5

© Pearson Education Limited 2010

30. Measurement 2	3 [F]	D, C, B	Measures and Construction: G3.4, G3.7
31. Enlargement and similarity	2 [F]	E, D, C, B	Properties of angles and shapes: G1.7, G1.7h, G1.8 Measures and Construction: G3.2
U1/U2 REVISION FOR MARCH RE-SITS (5 HOURS)			
32. Non-linear graphs	5 [F]	D, C, B	Expressions and Equations: N5.2h, N5.5h Sequences, Functions and Graphs: N6.7h, N6.8h, N6.11h, N6.13
33. Constructions and loci	5 [F]	C	Measures and Construction: G3.8, G3.10, G3.11
34. Pythagoras' theorem	6 [F]	C	Geometrical reasoning and calculation: G2.1
35. Trigonometry	0 [F]	B	Working with numbers and the number system: N1.14h Geometrical reasoning and calculation: G2.2h Measures and Construction: G3.6
36. Circle theorems	0 [F]	B	Properties of angles and shapes: G1.5h
UNIT 3 REVISION FOR JUNE EXAM (19 HOURS)			

Y11 SPRING TERM

Y11 SUMMER TERM

Detailed Modular Scheme of Work
begins on page 7

Longman AQA GCSE Maths Two-year Modular Scheme of Work starting with Unit 2 – For Middle sets taking Higher or Foundation Tier

Bold italic text indicates content that will only be examined at Higher Tier

Chapter 9 Number skills

Time: 4 hours [F]; 2 hours [H]

N1.2 Add, subtract, multiply and divide any number.

N1.4 Approximate to a given power of 10, up to three decimal places and one significant figure.

N1.4h Approximate to specified or appropriate degrees of accuracy, including a given number of decimal places and significant figures.

AQA Modular specification reference	Learning objectives	Grade	Resource	Common mistakes and misconceptions	Support and homework	
					Middle sets Teacher Guide	Middle sets Practice Book
			AQA GCSE Maths Middle sets Student Book; Middle sets Teacher Guide			
Number skills: adding and subtracting (N1.2); multiplying and dividing (N1.2)						
N1.2, N1.4	Multiply whole numbers using written methods Use repeated subtraction for division of whole numbers Round up or down in context	E, D	Section 9.1	Forgetting to add the numbers to find the final answer when using the grid method. Forgetting the zero when multiplying by tens when using the standard method. Writing 3.6 to represent 3 remainder 6. Not giving an answer in the context of the problem.	GPW 9.1	Section 9.1
N1.4, N1.4h	Check and estimate answers to problems Estimate answers to problems involving decimals Make estimates and approximations of calculations	E, D, C	Section 9.2	Finding an approximate value independent of the context in which it is set. Working out the actual answer instead of an approximation.		Section 9.2
N1.2	Multiply and divide negative numbers	E	Section 9.3	Applying the 'general rules' (*the signs are different, so the answer is negative; the signs are the same, so the answer is positive*) without constraint. Applying the rules for multiplying/dividing negative numbers to adding/subtracting negative numbers.		Section 9.3

© Pearson Education Limited 2010

Chapter 10 Factors, powers and standard form   Time: 3 hours [F]; 6 hours [H]

N1.6 The concepts and vocabulary of factor (divisor), multiple, common factor, highest common factor, least common multiple, prime number and prime factor decomposition.

N1.7 The terms square, positive and negative square root, cube and cube root.

N1.8 Index notation for squares, cubes and powers of 10.

N1.9 Index laws for multiplication and division of integer powers.

N1.9h Fractional and negative powers.

N1.10h Interpret, order and calculate numbers written in standard index form.

AQA Modular specification reference	Learning objectives	Grade	Resource AQA GCSE Maths Middle sets Student Book; Middle sets Teacher Guide	Common mistakes and misconceptions	Support and homework Middle sets Teacher Guide	Middle sets Practice Book
N1.6	Solve problems involving multiples Find lowest common multiples	E, C	Section 10.1	Confusing factors and multiples. Multiplying numbers with a common factor when attempting to find the LCM.	GPW 10.1/10.2	Section 10.1
N1.6	Solve problems involving factors Recognise two-digit prime numbers Find highest common factors	E, C	Section 10.2	Confusing factors and multiples. Missing out 1 as a factor. Forgetting whether it is the highest or lowest common factor that needs to be found. Thinking that 1 is a prime number.	GPW 10.1/10.2	Section 10.2
N1.7, N1.8	Calculate squares and cubes Calculate square roots and cube roots Understand the difference between positive and negative square roots Evaluate expressions involving squares, cubes and roots	E, D, C	Section 10.3	Multiplying by 2 instead of squaring. Writing $\sqrt{36} = -6$ or $\sqrt{36} = \pm 6$ when finding the negative square root.		Section 10.3
N1.8, N1.9h	*Understand and use index notation in calculations* *Understand and use negative powers and numbers to the power of 1 or 0*	*E, B*	*Section 10.4*	*Not being able to describe numbers written with powers.* *Working out 4^5 as 4×5.*	*GPW 10.4/10.6*	*Section 10.4*
N1.6	*Write a number as a product of prime factors using index notation* *Use prime factors to find HCFs and LCMs*	*C*	*Section 10.5*	*Not identifying the prime factors that appear in the decompositions of both numbers when finding the HCF.*	*GPW 10.5*	*Section 10.5*

| N1.9, N1.9h, N1.10h | Use laws of indices to multiply and divide numbers written in index notation

Carry out calculations with numbers given in standard form | C, B | Section 10.6 | Not converting the answers to calculations back into standard form (e.g. $32 \times 10^{-4} \Rightarrow 3.2 \times 10^{-3}$). | GPW
10.4/10.6 | Section
10.6 |

© Pearson Education Limited 2010

N4.1 Distinguish the different roles played by letter symbols in algebra, using the correct notation.

N5.1 Manipulate algebraic expressions by collecting like terms, by multiplying a single term over a bracket, and by taking out common factors.

N5.1h Multiply two linear expressions.

AQA Modular specification reference	Learning objectives	Grade	Resource AQA GCSE Maths Middle sets Student Book; Middle sets Teacher Guide	Common mistakes and misconceptions	Support and homework Middle sets Teacher Guide	Middle sets Practice Book
				Algebra skills: writing and simplifying expressions (N4.1, N5.1)		
N4.1, N5.1	Simplify algebraic expressions by collecting like terms	E	Section 11.1	Failing to comprehend that $x = 1x$. Combining unlike terms.		Section 11.1
N5.1	Multiply together two simple algebraic expressions	E	Section 11.2	Treating terms in m^2 and in m as like terms (e.g. simplifying $3m^2 + m$ wrongly to $4m^2$).		Section 11.2
N5.1	Multiply terms in a bracket by a number outside the bracket; Multiply terms in a bracket by a term that includes a letter	D	Section 11.3	Forgetting to multiply the second term in the bracket by the term outside (e.g. expanding $2(x + 3)$ as $2x + 3$), or ignoring minus signs (e.g. writing $3(m - 2)$ as $3m + 6$).	GPW 11.3	Section 11.3
N5.1	Simplify expressions involving brackets	D, C	Section 11.4	Forgetting to multiply the second term in the bracket by the term outside. Getting the wrong signs when multiplying negative values.		Section 11.4
N5.1	Recognise factors of algebraic terms; Simplify algebraic expressions by taking out common factors	D	Section 11.5	Not realising that x is a factor of x and x^2. Not taking out the highest common factor.	GPW 11.5	Section 11.5
N5.1h	*Multiply together two algebraic expressions with brackets; Square a linear expression*	*C, B*	*Section 11.6*	*Forgetting to multiply pairs of terms.*	*GPW 11.6*	*Section 11.6*

Chapter 12 Fractions

Time: 8 hours [F]; 6 hours [H]

N1.2 Add, subtract, multiply and divide any number.

N1.3 Understand and use number operations and the relationships between them, including inverse operations and hierarchy of operations.

N1.5 Order rational numbers.

N2.1 Understand equivalent fractions, simplifying a fraction by cancelling all common factors.

N2.2 Add and subtract fractions.

N2.7 Calculate with fractions, decimals and percentages.

AQA Modular specification reference	Learning objectives	Grade	Resource AQA GCSE Maths Middle sets Student Book; Middle sets Teacher Guide	Common mistakes and misconceptions	Support and homework Middle sets Teacher Guide	Middle sets Practice Book
N1.5, N2.1	Compare fractions with different denominators	E, D	Section 12.1	Multiplying the denominator but not the numerator when finding equivalent fractions.	GPW 12.1	Section 12.1
N2.2	Add and subtract fractions when one denominator is a multiple of the other Add and subtract fractions when both denominators have to be changed	E, D	Section 12.2	Adding/subtracting the denominators as well as the numerators. Not converting to equivalent fractions to make the denominators the same.		Section 12.2
N2.2	Add and subtract mixed numbers	C	Section 12.3	Incorrectly converting a mixed number to an improper fraction. Not converting the final answer back to a mixed number.	GPW 12.3	Section 12.3
N1.2	Multiply a fraction by a fraction	E	Section 12.4	Multiplying diagonally as though 'cross-multiplying' is being done (e.g. $\frac{2}{3} \times \frac{5}{6} = \frac{12}{15}$).		Section 12.4
N2.7	Multiply a whole number by a mixed number Multiply a fraction by a mixed number Multiply a mixed number by a mixed number	D, C, B	Section 12.5	Multiplying both the numerator and the denominator by the whole number (e.g. $3 \times \frac{5}{6} = \frac{15}{18}$).		Section 12.5
N1.3	Find the reciprocal of a whole number, a decimal or a fraction	C	Section 12.6	Leaving denominators as decimal numbers. Not simplifying answers when asked to do so.	GPW 12.6	Section 12.6

© Pearson Education Limited 2010

| N1.2 | Divide a whole number or a fraction by a fraction
Divide mixed numbers or fractions by whole numbers
Divide mixed numbers by mixed numbers | D, C, B | Section 12.7 | Finding the reciprocal of the wrong fraction, or finding the reciprocal of both fractions. | Section 12.7 |

N1.1 Understand integers and place value to deal with arbitrarily large positive numbers.

N1.2 Add, subtract, multiply and divide any number.

N2.3 Use decimal notation and recognise that each terminating decimal is a fraction.

N2.4 Recognise that recurring decimals are exact fractions, and that some exact fractions are recurring decimals.

AQA Modular specification reference	Learning objectives	Grade	Resource AQA GCSE Maths Middle sets Student Book; Middle sets Teacher Guide	Common mistakes and misconceptions	Support and homework	
					Middle sets Teacher Guide	Middle sets Practice Book
N1.1, N1.2	Add and subtract decimal numbers	E	Section 13.1	Not lining up the decimal points. Not recording the 'carry over' and forgetting to add it on. Not reducing a number during an exchange.	GPW 13.1	Section 13.1
N2.3	Convert decimals to fractions	D	Section 13.2	Working with the incorrect power of 10. Not giving answers in the simplest form.	GPW 13.2	Section 13.2
N1.2	Multiply and divide decimal numbers	D, C	Section 13.3	Working out the equivalent whole-number multiplication but forgetting to return to the decimal calculation at the end. Confusing multiplication with the rules for addition, writing a long multiplication with decimal points underneath each other.	GPW 13.3	Section 13.3
N2.3, N2.4	*Convert fractions to decimals Recognise recurring decimals Understand how recurring decimals relate to fractions*	*D, C, B*	*Section 13.4*	*Confusing 0.3 with $\frac{1}{3}$. Not understanding that recurring decimals are a form of exact maths and therefore rounding answers.*	*GPW 13.4*	*Section 13.4*

© Pearson Education Limited 2010

Chapter 14 Equations and inequalities

Time: 7 hours [F]; 7 hours [H]

N5.4 Set up and solve simple linear equations.

N5.4h Including simultaneous equations in two unknowns.

N5.7 Solve linear inequalities in one variable and represent the solution set on a number line.

N5.7h Solve linear inequalities in two variables, and represent the solution set on a suitable diagram.

AQA Modular specification reference	Learning objectives	Grade	Resource — AQA GCSE Maths Middle sets Student Book; Middle sets Teacher Guide	Common mistakes and misconceptions	Support and homework — Middle sets Teacher Guide	Support and homework — Middle sets Practice Book
N5.4	Solve two-step equations like $2x - 1 = 11$	E, D	Section 14.1	Not appreciating that an equation can be written in different but equivalent formats (e.g. $2a + 7 = 9 \rightarrow 7 + 2a = 9 \rightarrow 9 = 2a + 7$).	GPW 14.1a-14.4a GPW 14.1b-14.4b	Section 14.1
N5.4	Write and solve equations	E, D	Section 14.2	Not following a question carefully when writing an equation to represent a problem.	GPW 14.1a-14.4a GPW 14.1b-14.4b	Section 14.2
N5.4	Solve equations involving brackets	D, C	Section 14.3	Forgetting to multiply the second term in the bracket by the term outside. Getting the wrong signs when multiplying negative numbers. Incorrectly simplifying after expanding the bracket.	GPW 14.1a-14.4a GPW 14.1b-14.4b	Section 14.3
N5.4	Solve equations with an unknown on both sides	D, C	Section 14.4	Introducing errors when there are a negative number of unknowns on either side of the equation.	GPW 14.1a-14.4a GPW 14.1b-14.4b	Section 14.4
N5.4	*Solve equations involving fractions*	*C, B*	*Section 14.5*	*Incorrectly cancelling after multiplying by the LCM. Solving out of order (e.g. $\frac{x+2}{4} = 8$: trying to do −2 first).*		*Section 14.5*
N5.7, N5.7h	*Represent inequalities on a number line* *Write down whole-number values for unknowns in an inequality* *Solve inequalities*	*E, D, C, B*	*Section 14.6*	*Not reversing the sign when multiplying or dividing by a negative. Confusing the convention of an open circle for a strict inequality and a closed circle for an included boundary.*		*Section 14.6*
N5.4h	*Solve a pair of simultaneous equations*	*B*	*Section 14.7*	*Adding equations when they should be subtracted, and vice versa.*		*Section 14.7*

Chapter 15 Indices and formulae

Time: 6 hours [F]; 6 hours [H]

N1.9 Index laws for multiplication and division of integer powers.

N4.2 Distinguish in meaning between the words 'equation', 'formula', and 'expression'.

N5.6 Derive a formula, substitute numbers into a formula and change the subject of a formula.

AQA Modular specification reference	Learning objectives	Grade	Resource AQA GCSE Maths Middle sets Student Book; Middle sets Teacher Guide	Common mistakes and misconceptions	Support and homework Middle sets Teacher Guide	Middle sets Practice Book
N1.9	Use index notation in algebra Use index notation when multiplying or dividing algebraic terms	E, D, C	Section 15.1	Not realising that x means x^1, or that a number divided by 1 equals the number (e.g. $6 \div 1 = 6$).		Section 15.1
N1.9	Use index laws to multiply and divide powers in algebra Raise a number or variable to the power of 1 or 0 Use index laws for raising a power to another power	C, B	Section 15.2	Confusing unit and zero powers (e.g. stating that x means x^0, or that $x^0 = 0$).	GPW 15.2	Section 15.2
N4.2	Use algebra to write formulae in different situations	E, D	Section 15.3	Not seeing the 'general' case.		Section 15.3
N4.2, N5.6	Substitute numbers to work out the value of simple algebraic expressions Substitute numbers into expressions involving brackets and powers	E, D, C	Section 15.4	Incorrectly substituting values into expressions (e.g. substituting $a = 6$ into the expression $4a$, writing 46 and assuming it is forty-six). Ignoring BIDMAS.		Section 15.4
N4.2, N5.6	Substitute numbers into a variety of formulae	E, D	Section 15.5	Not realising that $\frac{n}{10}$ means $n \div 10$, or that $\frac{1}{2} \times 6$ means $\frac{1}{2}$ of $6 = 3$.	GPW 15.5	Section 15.5
N5.6	*Rearrange a formula to make a different variable the subject of the formula*	*C, B*	*Section 15.6*	*Not using brackets or a clear division (e.g. rewriting $c = 2a + 5$ as $a = c − 5 \div 2$).* *Not using the inverse operation (e.g. $x + y = z$ becomes $x = z + y$).*	*GPW 15.6*	*Section 15.6*

© Pearson Education Limited 2010

Chapter 16 Percentages

Time: 6 hours [F]; 5 hours [H]

N2.5 Understand that 'percentage' means 'number of parts per 100' and use this to compare proportions.

N2.7 Calculate with fractions, decimals and percentages.

N2.7h Including reverse percentage calculations.

AQA Modular specification reference	Learning objectives	Grade	Resource AQA GCSE Maths Middle sets Student Book; Middle sets Teacher Guide	Common mistakes and misconceptions	Support and homework Middle sets Teacher Guide	Middle sets Practice Book
	Number skills: fractions, decimals and percentages (N2.7); using percentages in calculations (N2.7)					
N2.7	Calculate a percentage increase or decrease	D	Section 16.1	Giving the actual increase/decrease as the answer when the amount after the increase/decrease is what is required. Using the multiplier as 1.5 rather than 1.05 for an increase of 5%. Writing '=' between quantities that are not equal, because the '=' sign is used as a shorthand for 'then I do this'.	GPW 16.1	Section 16.1
N2.5, N2.7	Perform calculations involving VAT Perform calculations involving credit Perform simple interest calculations	E, D	Section 16.2	Not seeing that 17.5% = 10% + 5% + 2.5%. Forgetting to add on the initial deposit in credit calculations.	GPW 16.2a, b	Section 16.2
N2.7	Calculate a percentage profit or loss	C	Section 16.3	Confusing cost price and selling price.	GPW 16.3	Section 16.3
N2.7	Perform calculations involving repeated percentage changes	C	Section 16.4	Not understanding when the multiplier should be greater than or less than 1. Using the multiplier as 1.5 rather than 1.05 for an increase of 5%.	GPW 16.4	Section 16.4
N2.7h	*Perform calculations involving finding the original quantity*	*B*	*Section 16.5*	*Not recognising that the problem is not a straightforward percentage increase/decrease question.* *Not using the correct multiplier.*	*GPW 16.5*	*Section 16.5*

Chapter 17 Sequences and proof

Time: 8 hours [F]; 6 hours [H]

N5.9 Use algebra to support and construct arguments.

N6.1 Generate terms of a sequence using term-to-term and position-to-term definitions of the sequence.

N6.2 Use linear expressions to describe the nth term of an arithmetic sequence.

AQA Modular specification reference	Learning objectives	Grade	Resource AQA GCSE Maths Middle sets Student Book; Middle sets Teacher Guide	Common mistakes and misconceptions	Support and homework Middle sets Teacher Guide	Middle sets Practice Book
N6.1	Find the next term in a sequence Describe the rule for continuing a sequence	E, D	Section 17.1	Expecting all sequences to have common differences. Looking at the first two numbers and assuming that the rest follow this pattern.		Section 17.1
N6.2	Find any term of a sequence given a formula for the nth term Find the nth term of a linear sequence	E, C	Section 17.2	Mistaking x^2 for $2x$.	GPW 17.2a, b	Section 17.2
N6.2	Find the nth term of a linear sequence Use the nth term to find terms in a sequence	C	Section 17.3	Not comprehending that when finding the 50th term it is not necessary to find all the terms up to that term.		Section 17.3
N6.2	Find the next few terms in a sequence of patterns Find the nth term for a sequence of diagrams	E, C	Section 17.4	Not writing down the sequence and trying to do it all mentally. Not making the connection between the structure of the physical pattern and the form the nth term takes.		Section 17.4
N6.1	Find the first few terms of a quadratic sequence by using the nth term Find the next few terms of a quadratic sequence by looking at differences	D	Section 17.5	Mistaking x^2 for $2x$.		Section 17.5
N6.1	Find the nth term of a simple quadratic sequence	C	Section 17.6	Thinking that $(3x)^2 = 3x^2$. Not checking whether the derived quadratic rule works.		Section 17.6
N5.9	Show step-by-step deduction when proving results	E, D, C	Section 17.7	Not appreciating that a proof shows something works for all values.	GPW 17.7	Section 17.7
N5.9	Show something is false by using a counter-example	C	Section 17.8	Assuming that 'number' means positive whole number. Not identifying an appropriate counter-example.	GPW 17.8	Section 17.8

© Pearson Education Limited 2010

Chapter 7 Ratio and proportion

N3.1 Use ratio notation, including reduction to its simplest form and its various links to fraction notation.

N3.2 Divide a quantity in a given ratio.

N3.3 Solve problems involving ratio and proportion, including the unitary method of solution.

AQA Modular specification reference	Learning objectives	Grade	Resource AQA GCSE Maths Middle sets Student Book; Middle sets Teacher Guide	Common mistakes and misconceptions	Support and homework	
					Middle sets Teacher Guide	Middle sets Practice Book
N3.1, N3.2	Simplify a ratio to its lowest terms Use a ratio in practical situations	E, D	Section 7.1	Swapping over the numbers in the ratio (e.g. 2 : 5 becomes 5 : 2). Simplifying ratios without ensuring the quantities are in the same units.	GPW 7.1a, b	Section 7.1
N3.1, N3.3	Write a ratio as a fraction Use a ratio to find one quantity when the other is known	D, C	Section 7.2	Turning a ratio into a fraction (e.g. the ratio 4 : 5 becomes $\frac{4}{5}$). Failing to find the value of the unit fraction in more complex problems.	GPW 7.2	Section 7.2
N3.3	Write a ratio in the form 1 : n or n : 1	C	Section 7.3	Ignoring different units in a ratio (e.g. simplifying 2 days : 15 hours to 1 : $7\frac{1}{2}$) .	GPW 7.3	Section 7.3
N3.3	Share a quantity in a given ratio	D, C	Section 7.4	Forgetting to work out the total number of parts first. Using a ratio as a fraction.	GPW 7.4	Section 7.4
N3.3	Solve word problems involving ratio	C	Section 7.5	Not multiplying both sides of the ratio by the same number. Giving an answer without considering the context.		Section 7.5
N3.3	Understand direct proportion Solve proportion problems, including using the unitary method	D	Section 7.6	Not always seeing the relationships between numbers (e.g. if the cost of 4 items is given, and the price of 8 is asked for).		Section 7.6
N3.3	Work out which product is the better buy	D	Section 7.7	Not making the units the same for each item. Comparing unlike unit rates (e.g. price per gram for one item but amount for 1p for the other).	GPW 7.7	Section 7.7
N3.3	*Solve word problems involving direct and inverse proportion* *Understand inverse proportion*	*D, C, B*	*Section 7.8*	*Dividing by the wrong quantity in conversion problems.*	*GPW 7.8*	*Section 7.8*

Chapter 18 Linear graphs

Time: 7 hours [F]; 7 hours [H]

N5.4h Including simultaneous equations in two unknowns.

N5.7h Solve linear inequalities in two variables, and represent the solution set on a suitable diagram.

N6.3 Use the conventions for coordinates in the plane and plot points in all four quadrants, including using geometric information.

N6.4 Recognise and plot equations that correspond to straight-line graphs in the coordinate plane, including finding their gradients.

N6.5h Understand that the form $y = mx + c$ represents a straight line and that m is the gradient of the line and that c is the value of the y-intercept.

N6.6h Understand the gradients of parallel lines.

N6.11 Construct linear functions from real-life problems and plot their corresponding graphs.

N6.12 Discuss, plot and interpret graphs (which may be non-linear) modelling real situations.

AQA Modular specification reference	Learning objectives	Grade	Resource — AQA GCSE Maths Middle sets Student Book; Middle sets Teacher Guide	Common mistakes and misconceptions	Support and homework — Middle sets Teacher Guide	Middle sets Practice Book
N6.3	Find the mid-point of a line segment	D, C	Section 18.1	Subtracting the coordinates (instead of calculating an average) when finding the mid-point.		Section 18.1
N6.4	Recognise straight-line graphs parallel to the x- or y-axis / Work out coordinates of points of intersection when two graphs cross / Plot graphs of linear functions	E, D, C	Section 18.2	Incorrectly calibrating the coordinate axes. Not using a third point as a check when drawing a straight line.	GPW 18.2	Section 18.2
N6.5h, N6.6h	Plot straight-line graphs / Find the gradient of a straight-line graph / Understand the meaning of m and c in the equation $y = mx + c$ / Find the equation of a line	D, C, B	Section 18.3	Forgetting the negative on the gradient.	GPW 18.3	Section 18.3
N6.11	Plot and use conversion graphs	E	Section 18.4	Inaccurately reading from one value on a conversion graph to find another value.		Section 18.4
N6.11, N6.12	Draw, read and interpret distance–time graphs / Sketch and interpret real-life graphs	E, D, C	Section 18.5	Drawing and labelling axes before working out the axes range appropriate to the problem.		Section 18.5
N5.4h	*Use a graphical method to solve simultaneous equations*	*B*	*Section 18.6*	*Forgetting to ensure that the individual lines are drawn accurately.* *Not appreciating that the graphical solutions to simultaneous equations are only approximate.*	*GPW 18.6*	*Section 18.6*
N5.7h	*Solve inequalities graphically*	*B*	*Section 18.7*	*Mixing up whether the lines should be dotted or solid.* *Shading the incorrect area.*		*Section 18.7*

© Pearson Education Limited 2010

Chapter 19 Quadratic equations

N5.2h Factorise quadratic expressions, including the difference of two squares.

N5.5h Solve quadratic equations.

AQA Modular specification reference	Learning objectives	Grade	Resource AQA GCSE Maths Middle sets Student Book; Middle sets Teacher Guide	Common mistakes and misconceptions	Support and homework Middle sets Teacher Guide	Middle sets Practice Book
N5.2h	Factorise a quadratic expression that is the difference of two squares	B	Section 19.1	Incorrectly expanding brackets (e.g. expanding $(x-3)^2$ as x^2-9).		Section 19.1
N5.2h	Factorise a quadratic of the form $x^2 + bx + c$	B	Section 19.2	Looking for numbers whose sum is c and product b.	GPW 19.2	Section 19.2
N5.2h, N5.5h	Solve quadratic equations by rearranging Solve quadratic equations by factorising	B	Section 19.3	Forgetting that there are two solutions to a quadratic equation.	GPW 19.3	Section 19.3
N5.2h, N5.5h	Write quadratic equations for problems and then solve them	B	Section 19.4	Not identifying which is the unknown value, or using two variables.		Section 19.4

End of Unit 2

Chapter 1 Data collection

Time: 6 hours [F]; 5 hours [H]

S1 Understand and use the statistical problem solving process which involves

- specifying the problem and planning
- processing and presenting the data
- collecting data
- interpreting and discussing the results.

S2.1 Types of data: qualitative, discrete, continuous. Use of grouped and ungrouped data.

S2.2 Identify possible sources of bias.

S2.3 Design an experiment or survey.

S2.4 Design data collection sheets distinguishing between different types of data.

S2.5 Extract data from printed tables and lists.

S3.1 Design and use two-way tables for grouped and ungrouped data.

AQA Modular specification reference	Learning objectives	Grade	Resource AQA GCSE Maths Middle sets Student Book; Middle sets Teacher Guide	Common mistakes and misconceptions	Support and homework Middle sets Teacher Guide	Middle sets Practice Book
S1	Learn about the data handling cycle Know how to write a hypothesis	D	Section 1.1	Formulating a hypothesis that cannot be tested. Thinking that a hypothesis is not valuable if it is eventually proved false.		Section 1.1
S2.3, S2.4	Know where to look for information	D	Section 1.2	Not realising that data collected by a third party (even if the results of a survey or experiment) is classed as secondary data.		Section 1.2
S2.1	Be able to identify different types of data	D	Section 1.3	Not appreciating that some data can be treated as either discrete or continuous depending on the context (e.g. age – this is really continuous, but is often treated as discrete, such as when buying child or adult tickets).		Section 1.3
S2.4	Work out methods for gathering data efficiently	E	Section 1.4	Using shortcuts in the tallying process – counting up the items in each class, rather than tallying items one by one.		Section 1.4
S2.4, S2.5	Work out methods for gathering data that can take a wide range of values	D	Section 1.5	Using overlapping class intervals. Recording data which is on the boundary of a class interval in the wrong class.		Section 1.5
S2.5, S3.1	Work out methods for recording related data	D	Section 1.6	Not checking that the totals in two-way tables add up.		Section 1.6
S2.3, S2.4	Learn how to write good questions to find out information	C	Section 1.7	Using overlapping classes, or gaps between classes, for response options.		Section 1.7
S2.2, S2.3, S2.4	Know the techniques to use to get a reliable sample	C	Section 1.8	Mistaking biased samples for random samples.		Section 1.8

© Pearson Education Limited 2010

Chapter 2 Fractions, decimals and percentages

Time: 6 hours [F]; 4 hours [H]

N1.14 Use calculators effectively and efficiently, including statistical functions.

N2.6 Interpret fractions, decimals and percentages as operators.

N2.7 Calculate with fractions, decimals and percentages.

AQA Modular specification reference	Learning objectives	Grade	Resource AQA GCSE Maths Middle sets Student Book; Middle sets Teacher Guide	Common mistakes and misconceptions	Support and homework Middle sets Teacher Guide	Middle sets Practice Book
N2.7	Find a fraction of an amount with a calculator Find a fraction of an amount with a calculator in more complex situations	E, D	Section 2.1	Incorrectly inputting numbers on the calculator. Being unsure of what to work out when a fraction calculation is set in context.	GPW 2.1	Section 2.1
N2.7	Write one quantity as a fraction of another	D	Section 2.2	Not making the denominator the total in questions involving a number of quantities. Working with quantities in different units. Incorrectly cancelling down.	GPW 2.2	Section 2.2
N1.14	Use the fraction key on a calculator Use the fraction key on a calculator with mixed numbers	E, D	Section 2.3	Not recognising or know how to use the fraction key on a calculator. Misinterpreting a mixed number on a calculator display.		Section 2.3
N2.7	Find a percentage of an amount with a calculator Find percentages of amounts in more complex situations	E, D	Section 2.4	Thinking that percentages over 100% cannot exist. Treating a percentage such as 0.05% as though it were 5%. Adding the percentage to the cost when finding a percentage increase (e.g. £315 + 15% VAT = £330).		Section 2.4
N2.7	Write one quantity as a percentage of another Write one quantity as a percentage of another in more complex situations	D, C	Section 2.5	Not using the original amount as the denominator, when finding a percentage difference. Working with quantities in different units.		Section 2.5
N2.6, N2.7	Calculate a percentage increase or decrease	D	Section 2.6	Giving the actual increase/decrease as the answer when the amount after the increase/decrease is what is required. Using the multiplier as 1.5 rather than 1.05 for an increase of 5%.		Section 2.6
N2.7	Understand and use a retail prices index Understand and use a retail prices index in more complex situations	D, C	Section 2.7	Using a previously found price instead of the base year price.	GPW 2.7	Section 2.7

Chapter 3 Interpreting and representing data

Time: 8 hours [F]; 6 hours [H]

S3.2 Produce charts and diagrams for various data types. Scatter graphs, stem-and-leaf, tally charts, pictograms, bar charts, dual bar charts, pie charts, line graphs, frequency polygons, histograms with equal class intervals.

S4.2 Look at data to find patterns and exceptions.

S4.3 Recognise correlation and draw and/or use lines of best fit by eye, understanding what they represent.

AQA Modular specification reference	Learning objectives	Grade	Resource: AQA GCSE Maths Middle sets Student Book; Middle sets Teacher Guide	Common mistakes and misconceptions	Support and homework: Middle sets Teacher Guide	Support and homework: Middle sets Practice Book
S3.2	Draw a pie chart	E	Section 3.1	Not drawing the angles in the pie chart accurately or using the appropriate scale on the protractor. Measuring each angle from the same starting point.	GPW 3.1	Section 3.1
S3.2	Draw stem-and-leaf diagrams	D	Section 3.2	Forgetting to put a key and order the leaves. Forgetting to recombine the stem and leaf and just giving the leaf as the value.		Section 3.2
S3.2, S4.2	Draw a scatter diagram on a given grid; Interpret points on a scatter diagram	D	Section 3.3	Assuming that all the plotted points must be joined with a line. Drawing the diagram without spending time working out the best scale.		Section 3.3
S4.3	Draw a line of best fit on a scatter diagram; Describe types of correlation; Use the line of best fit	D, C	Section 3.4	Trying to make the line of best fit go through the origin, rather than drawing it appropriately. Not appreciating correlation in terms of 'positive' and 'negative'.		Section 3.4
S3.2	Draw frequency diagrams for grouped data	D	Section 3.5	Using grouped labels on the data axes (e.g. 15–20, rather than the ends of the bar being clearly marked with a 15 at one end and a 20 at the other end).		Section 3.5
S3.2	Draw frequency polygons for grouped data	C	Section 3.6	Using a grouped label on the horizontal axis rather than a continuous scale. Plotting the upper bound instead of the mid-point.	GPW 3.6	Section 3.6

© Pearson Education Limited 2010

Chapter 4 Range and averages

Time: 6 hours [F]; 4 hours [H]

S3.3 Calculate median, mean, range, mode and modal class.

S4.1 Interpret a wide range of graphs and diagrams and draw conclusions.

AQA Modular specification reference	Learning objectives	Grade	Resource AQA GCSE Maths Middle sets Student Book; Middle sets Teacher Guide	Common mistakes and misconceptions	Support and homework Middle sets Teacher Guide	Middle sets Practice Book
S3.3, S4.1	Find the mean, median and range from a set of data, including data given in a stem and leaf diagram	E, D	Section 4.1	Omitting units when writing averages or range. Forgetting to include the stem when reading the median from a stem and leaf diagram.	GPW 4.1	Section 4.1
S3.3, S4.1	Calculate the mode, median and range from an ungrouped frequency table	E	Section 4.2	Confusing the frequencies and the data values.		Section 4.2
S3.3, S4.1	Calculate the mean from an ungrouped frequency table	D, C	Section 4.3	Dividing by the number of rows in the frequency table (i.e. the number of different data values), not by the sum of the frequencies.		Section 4.3
S3.3, S4.1	Find the modal class from a grouped frequency table Estimate the range from a grouped frequency table Work out the class interval which contains the median from data given in a grouped frequency table	D, C	Section 4.4	Not appreciating that the statistics calculated from grouped frequency tables are estimates. Not understanding that the estimate for the range is an upper limit.	GPW 4.4/4.5	Section 4.4
S3.3, S4.1	Estimate the mean of data given in a grouped frequency table	C	Section 4.5	Incorrectly calculating the mid-points of class intervals for grouped discrete data (e.g. the mid-point of the class interval 10–19 is 14.5, not 15). Interpreting 'find an estimate for the mean' as 'guess the mean'.	GPW 4.4/4.5	Section 4.5

Chapter 5 Probability

Time: 8 hours [F]; 7 hours [H]

S3.1 Design and use two-way tables for grouped and ungrouped data.

S5.1 Understand and use the vocabulary of probability and the probability scale.

S5.2 Understand and use estimates or measures of probability from theoretical models (including equally likely outcomes), or from relative frequency.

S5.3 List all outcomes for single events, and for two successive events, in a systematic way and derive related probabilities.

S5.4 Identify different mutually exclusive outcomes and know that the sum of the probabilities of all these outcomes is 1.

S5.5h Know when to add or multiply two probabilities: if A and B are mutually exclusive, then the probability of A or B occurring is P(A) + P(B), whereas if A and B are independent events, the probability of A and B occurring is P(A) × P(B).

S5.6h Use tree diagrams to represent outcomes of compound events, recognising when events are independent.

S5.7 Compare experimental data and theoretical probabilities.

S5.8 Understand that if an experiment is repeated, this may – and usually will – result in different outcomes.

S5.9 Understand that increasing sample size generally leads to better estimates of probability and population characteristics.

AQA Modular specification reference	Learning objectives	Grade	Resource AQA GCSE Maths Middle sets Student Book; Middle sets Teacher Guide	Common mistakes and misconceptions	Support and homework Middle sets Teacher Guide	Middle sets Practice Book
S5.1, S5.4	Work out the probability of an event not happening when you know the probability that it does happen	E	Section 5.1	Not remembering that probabilities can be written as fractions, decimals and percentages. Incorrectly subtracting decimals from 1.		Section 5.1
S5.4	Understand and use the fact that the sum of the probabilities of all mutually exclusive outcomes is 1	D	Section 5.2	Adding or subtracting the incorrect values due to misreading the question.	GPW 5.2	Section 5.2
S3.1, S5.3	Understand and use two-way tables	E, D	Section 5.3	Giving an answer that is not requested. Reading the data from the table incorrectly.		Section 5.3
S5.2	Predict the likely number of successful events given the probability of any outcome and the number of trials or experiments	D	Section 5.4	Incorrectly finding fractions of an amount. Not cancelling a fraction to its simplest form.	GPW 5.4	Section 5.4
S5.2, S5.7, S5.8, S5.9	Estimate probabilities from experimental data	C	Section 5.5	Trying to plot decimals worked out to three decimal places or more. Comparing theoretical probability with relative frequency without taking into account the number of trials carried out.		Section 5.5
S5.5h	Calculate the probability of two independent events happening at the same time	C	Section 5.6	Not recognising when a question involves independent events and so adding rather than multiplying the fractions.	GPW 5.6	Section 5.6

© Pearson Education Limited 2010

25

| S5.6h | Use and understand tree diagrams in simple contexts | B | Section 5.7 | Adding probabilities along the branch rather than multiplying. | Section 5.7 |

Chapter 6 Cumulative frequency

Time: 0 hours [F]; 6 hours [H]

S3.2h Histograms with equal or unequal class intervals, box plots, cumulative frequency diagrams, relative frequency diagrams.

S3.3h Quartiles and inter-quartile range.

S4.4 Compare distributions and make inferences.

AQA Modular specification reference	Learning objectives	Grade	Resource AQA GCSE Maths Middle sets Student Book; Middle sets Teacher Guide	Common mistakes and misconceptions	Support and homework	
					Middle sets Teacher Guide	Middle sets Practice Book
S3.2h, S3.3h	Compile a cumulative frequency table and draw cumulative frequency diagrams Use cumulative frequency diagrams to analyse data	B	Section 6.1	Mis-reading the graph axes scales. Inaccurately summing the frequencies.		Section 6.1
S3.2h	Draw a box plot from a cumulative frequency diagram	B	Section 6.2	Mis-reading the medians and quartiles. Drawing the lower end of the box plot at zero, rather than at the bottom of the lowest class.		Section 6.2
S3.2h, S4.4	Use cumulative frequency diagrams and box plots to compare data and draw conclusions	B	Section 6.3	Inaccurately plotting cumulative frequency diagrams and box plots. Not appreciating the need for a coherent written analysis of diagrams.		Section 6.3

Chapter 7 Ratio and proportion is taught in Unit 2

27

© Pearson Education Limited 2010

N1.10h Interpret, order and calculate numbers written in standard index form.

N2.7h Including reverse percentage calculations.

AQA Modular specification reference	Learning objectives	Grade	Resource AQA GCSE Maths Middle sets Student Book; Middle sets Teacher Guide	Common mistakes and misconceptions	Support and homework Middle sets Teacher Guide	Middle sets Practice Book
N2.7h	Perform calculations involving repeated percentage changes	C	Section 8.1	Leaving the multiplier as a percentage, instead of converting to a decimal. Inaccurately converting to a decimal. Not understanding compounding (e.g. treating compound interest as simple).	GPW 8.1	Section 8.1
N2.7h	*Perform reverse percentage calculations*	*B*	*Section 8.2*	*Incorrectly converting percentages to decimals (e.g. 5% = 0.5). Failing to add an increase or subtract a decrease to the given quantity.*		*Section 8.2*
N1.10h	*Interpret and use standard form*	*B*	*Section 8.3*	*Inaccurately converting from a factor of 10 to 10^x and vice versa. Forgetting to include the minus signs in the power for numbers less than 1. Incorrectly entering a number in standard form on a calculator.*		*Section 8.3*

End of Unit 1

Chapter 20 Number skills revisited

Time: 3 hours [F]; 2 hour [H]

N1.3 Understand and use number operations and the relationships between them, including inverse operations and hierarchy of operations.

N1.4 Approximate to a given power of 10, up to three decimal places and one significant figure.

N1.14 Use calculators effectively and efficiently.

N2.1 Understand equivalent fractions, simplifying a fraction by cancelling all common factors.

N2.5 Understand that 'percentage' means 'number of parts per 100' and use this to compare proportions.

N2.7 Calculate with fractions, decimals and percentages.

N3.1 Use ratio notation, including reduction to its simplest form and its various links to fraction notation.

AQA Modular specification reference	Learning objectives	Grade	Resource	Common mistakes and misconceptions	Support and homework	
			AQA GCSE Maths Middle sets Student Book; Middle sets Teacher Guide		Middle sets Teacher Guide	Middle sets Practice Book
N1.3, N1.4, N1.14, N2.1, N2.5, N2.7, N3.1	Understand equivalent fractions		Chapter 20	Forgetting to multiply/divide both the numerator and denominator when finding equivalent fractions and simplifying fractions.		Chapter 20
	Simplify a fraction by cancelling all common factors					
	Recognise that each terminating decimal is a fraction			Applying an incorrect understanding of 'reciprocal'.		
	Convert simple fractions to percentages and vice versa			Writing the ratio in the incorrect order.		
	Use percentages to compare proportions			Forgetting to use BIDMAS when using calculators to perform calculations.		
	Understand 'reciprocal' as multiplicative inverse			Not giving an answer in the context of the problem.		
	Use ratio notation			Treating the digits each side of the decimal point as separate whole numbers , so that 0.95 rounded to 1 d.p. = 0.1.		
	Use brackets and the hierarchy of operations					
	Add, subtract, multiply and divide integers			Dropping zeros when rounding to a number of significant figures (e.g. 5840 = 6 to 1 s.f.).		
	Use calculators effectively and efficiently; use function keys for squares					
	Use inverse operations					
	Round to the nearest integer, to one significant figure and to one, two or three decimal places					
	Give solutions in the context of the problem to an appropriate degree of accuracy					

© Pearson Education Limited 2010

Chapter 21 Angles

Time: 5 hours [F]; 3 hours [H]

G1.1 Recall and use properties of angles at a point, angles at a point on a straight line (including right angles), perpendicular lines, and opposite angles at a vertex.

G1.2 Understand and use the angle properties of parallel and intersecting lines, triangles and quadrilaterals.

G3.1 Use and interpret maps and scale drawings.

G3.6 Understand and use bearings.

AQA Modular specification reference	Learning objectives	Grade	Resource **AQA GCSE Maths Middle sets Student Book; Middle sets Teacher Guide**	Common mistakes and misconceptions	Support and homework **Middle sets Teacher Guide**	**Middle sets Practice Book**
			Geometry skills: angles (G1.1)			
G1.1	Calculate angles around a point Recognise vertically opposite angles	E	Section 21.1	Measuring rather than calculating angles.		Section 21.1
G1.2	Recognise corresponding and alternate angles Calculate angles in diagrams with parallel lines	D	Section 21.2	Confusing alternate and corresponding angles.	GPW 21.2	Section 21.2
G3.1, G3.6	Use three-figure bearing notation Measure the bearing from one place to another Plot a bearing Calculate bearings for return journeys Draw and interpret scale diagrams to represent journeys	E, D, C	Section 21.3	Confusing which angles need to be found. Not realising that some of the angles asked for can simply be read off the diagram.	GPW 21.3	Section 21.3

Chapter 22 Measurement 1

Time: 4 hours [F]; 3 hours [H]

N1.3 Understand and use number operations and the relationships between them, including inverse operations and hierarchy of operations.

G3.1 Use and interpret maps and scale drawings.

G3.3 Interpret scales on a range of measuring instruments and recognise the inaccuracy of measurements.

G3.4 Convert measurements from one unit to another.

AQA Modular specification reference	Learning objectives	Grade	Resource AQA GCSE Maths Middle sets Student Book; Middle sets Teacher Guide	Common mistakes and misconceptions	Support and homework Middle sets Teacher Guide	Middle sets Practice Book
N1.3	Solve problems involving times, dates and timetables	E	Section 22.1	Confusing the decimal parts of an hour with hours and minutes (e.g. writing 1.25 hours as 1 hour 25 minutes) and vice versa.	GPW 22.1	Section 22.1
G3.4	Know and use approximate metric equivalents of pounds, feet, miles, pints and gallons	E	Section 22.2	Not considering the relative size of units when deciding whether to multiply or divide.	GPW 22.2	Section 22.2
G3.1	Use and interpret maps and scale drawings	E	Section 22.3	Missing out steps when converting between (for example) km and cm. Not making allowances when measurements are given in a variety of units.		Section 22.3
G3.3	Recognise that measurements given to the nearest whole unit may be inaccurate by up to one half unit in either direction	C	Section 22.4	Difficulty comprehending the definition of the upper bound, since, for example, 146.5 rounds to 147.		Section 22.4

© Pearson Education Limited 2010

Chapter 23 Triangles and constructions

Time: 4 hours [F]; 3 hours [H]

G1.1 Recall and use properties of angles at a point, angles at a point on a straight line (including right angles), perpendicular lines, and opposite angles at a vertex.

G1.2 Understand and use the angle properties of parallel and intersecting lines, triangles and quadrilaterals.

G1.8 Understand congruence and similarity.

G3.9 Draw triangles and other 2D shapes using a ruler and protractor.

G3.10 Use straight edge and a pair of compasses to do constructions.

AQA Modular specification reference	Learning objectives	Grade	Resource AQA GCSE Maths Middle sets Student Book; Middle sets Teacher Guide	Common mistakes and misconceptions	Support and homework	
					Middle sets Teacher Guide	Middle sets Practice Book
G1.1, G1.2	Solve angle problems in triangles Solve angle problems in triangles involving algebra	E, D	Section 23.1	Not realising when a triangle is isosceles and thinking that the problem cannot be solved. Trying to do too many steps in one go when answering algebra-based questions.	GPW 23.1	Section 23.1
G3.9, G3.10	Draw triangles accurately when given the length of all three sides Draw triangles accurately when at least one angle is given	E, D	Section 23.2	Inaccurately using a protractor or compasses. Not completing the triangle by drawing the third side. Rubbing out construction lines.		Section 23.2
G1.8	Recognise and explain how triangles are congruent	C	Section 23.3	Thinking that two triangles are congruent when they are not (due to the relative positions of side lengths or angles being in different positions). Incorrectly assuming that by showing that the three sets of angles are the same in both triangles congruency is proved.		Section 23.3

Chapter 24 Equations, formulae and proof

N4.2 Distinguish in meaning between the words 'equation', 'formula', and 'expression'.

N5.1 Manipulate algebraic expressions by collecting like terms, by multiplying a single term over a bracket, and by taking out common factors.

N5.4 Set up and solve simple linear equations.

N5.6 Derive a formula, substitute numbers into a formula.

N5.8 Use systematic trail and improvement to find approximate solutions of equations where there is no simple analytical method of solving them.

G2.3 Justify simple geometrical properties.

G2.3h Simple geometrical proofs.

AQA Modular specification reference	Learning objectives	Grade	Resource AQA GCSE Maths Middle sets Student Book; Middle sets Teacher Guide	Common mistakes and misconceptions	Support and homework Middle sets Teacher Guide	Middle sets Practice Book
Algebra skills: expressions (N4.2, N5.1); brackets (N5.1); solving equations (N5.1); substituting into a formula to solve problems						
N5.4, N5.6	Write your own formulae and equations Substitute into a formula to solve problems Set up and solve equations Change the subject of a formula	D, C, B	Section 24.1	Failing to consider the different terms of an expression when changing the subject of a formula (e.g. $W = \frac{1}{2}x + 3 \Rightarrow 2W = x + 3$). Not using brackets or a clear division (e.g. rewriting $c = 2a + 5$ as $a = c - 5 \div 2$). Not using the inverse operation (e.g. $x + y = z$ becomes $x = z + y$).	GPW 24.1	Section 24.1
G2.3, G2.3h	Prove simple results from geometry	C	Section 24.2	Not laying out answers in an organised way. Not providing reasons for each stage of the working.		Section 24.2
N5.8	Use trial and improvement to find solutions to equations	C	Section 24.3	Not checking the mid-point to determine which of two values is correct (e.g. choosing between $x = 3.3$ and $x = 3.4$ based on the value of the function and the desired output). Using the value of the equation as the answer rather than the value of the variable.	GPW 24.3	Section 24.3

© Pearson Education Limited 2010

N5.4 Set up and solve simple linear equations.

N6.3 Use the conventions for coordinates in the plane and plot points in all four quadrants, including geometric information.

G1.2 Understand and use the angle properties of parallel and intersecting lines, triangles and quadrilaterals.

G1.3 Calculate and use the sums of the interior and exterior angles of polygons.

G1.4 Recall the properties and definitions of special types of quadrilateral, including square, rectangle, parallelogram, trapezium, kite and rhombus.

AQA Modular specification reference	Learning objectives	Grade	Resource **AQA GCSE Maths Middle sets Student Book; Middle sets Teacher Guide**	Common mistakes and misconceptions	Support and homework **Middle sets Teacher Guide**	**Middle sets Practice Book**
G1.2, N5.4	Calculate the interior angles of quadrilaterals Solve angle problems in quadrilaterals involving algebra	E, D	Section 25.1	Working things out mentally without writing down the calculations.	GPW 25.1	Section 25.1
G1.2, G1.4	Make quadrilaterals from two triangles Use parallel lines and other angle properties in quadrilaterals	E, D	Section 25.2	Giving correct answers but not explaining the properties used.		Section 25.2
G1.3, N5.4	Use the exterior angles of polygons to solve problems Solve more complex angle problems involving exterior and interior angles of a polygon	D, C	Section 25.3	Incorrectly splitting the polygon into triangles.	GPW 25.3a, b	Section 25.3
N6.3	Plot all points of a quadrilateral given geometric information Find the mid-point of a line plotted on the coordinate axes	E	Section 25.4	Plotting the numbers on the x- and y-axes the wrong way round. Not recognising, or be able to name, some of the less common quadrilaterals (e.g. the kite and trapezium). Averaging only the x- or y-coordinate and not both when finding the mid-point.		Section 25.4

Chapter 26 Perimeter, area and volume

G4.1 Calculate perimeters and areas of shapes made from triangles and rectangles.

G4.4 Calculate volumes of right prisms and of shapes made from cubes and cuboids.

AQA Modular specification reference	Learning objectives	Grade	Resource **AQA GCSE Maths Middle sets Student Book; Middle sets Teacher Guide**	Common mistakes and misconceptions	Support and homework **Middle sets Teacher Guide**	**Middle sets Practice Book**
G4.1	Find the perimeter and area of rectangles, parallelograms, triangles and trapezia	E, D	Section 26.1	Not making rough estimates of areas as a check to avoid arithmetical errors. Incorrectly converting between units. Using measurements in different units.	GPW 26.1	Section 26.1
G4.1	Find the perimeter and area of compound shapes	E, D	Section 26.2	Incorrectly calculating missing lengths. Adding areas instead of subtracting.		Section 26.2
G4.1, G4.4	Find the volume and surface area of a prism	E, D, C	Section 26.3	Confusing volume and surface area.	GPW 26.3	Section 26.3

© Pearson Education Limited 2010

G2.4 Use 2D representations of 3D shapes.

	Learning objectives	Grade	Resource	Common mistakes and misconceptions	Support and homework	
					Middle sets Teacher Guide	Middle sets Practice Book
AQA Modular specification reference			**AQA GCSE Maths Middle sets Student Book; Middle sets Teacher Guide**			
G2.4	Make a drawing of a 3-D object on isometric paper Draw plans and elevations of 3-D objects Identify planes of symmetry of 3-D objects	E, D	Section 27.1	Missing out hidden cubes when converting from a 3-D view to a plan or elevation. Using isometric paper in landscape not in portrait.		Section 27.1

G1.6 Recognise reflection and rotation symmetry of 2D shapes.

G1.7 Describe and transform 2D shapes using single or combined rotations, reflections, translations, or enlargements by a positive scale factor and distinguish properties that are preserved under particular transformations.

G5.1 Understand and use vector notation for translations.

AQA Modular specification reference	Learning objectives	Grade	Resource AQA GCSE Maths Middle sets Student Book; Middle sets Teacher Guide	Common mistakes and misconceptions	Support and homework	
					Middle sets Teacher Guide	Middle sets Practice Book
G1.7	Draw reflections on a coordinate grid Describe reflections on a coordinate grid	E, D, C	Section 28.1	Drawing the image a different distance from the mirror line than the object. Incorrectly identifying mirror lines parallel to the *x*- or *y*-axis.	GPW 28.1	Section 28.1
G1.7, G5.1	Translate a shape on a grid Use column vectors to describe translations	D, C	Section 28.2	Forgetting what the two values in the column vector mean. Using coordinate notation instead of vector notation. Confusing the terms 'transformation' and 'translation'.		Section 28.2
G1.6, G1.7	Draw the position of a shape after rotation about a centre Describe a rotation fully giving the size and direction of turn and the centre of rotation	D, C	Section 28.3	Working out the angle of rotation incorrectly. Turning in the wrong direction. Ignoring the centre of rotation when it is outside the shape.	GPW 28.3	Section 28.3
G1.7	*Transform shapes using more than one transformation* *Describe combined transformations of shapes on a grid*	*C*	*Section 28.4*	*Not appreciating that two transformations, one followed by another, may be equivalent to a single transformation.* *Not understanding that for 'general' questions where no shape is given, transformations can be carried out on different shapes to investigate what happens.*		*Section 28.4*

© Pearson Education Limited 2010

Chapter 29 Circles and cylinders

G1.5 Distinguish between centre, radius, chord, diameter, circumference, tangent, arc, sector and segment.

G2.4 Use 2D representations of 3D shapes.

G4.1 Calculate perimeters and areas of shapes made from triangles and rectangles.

G4.3 Calculate circumferences and areas of circles.

G4.4 Calculate volumes of right prisms and of shapes made from cubes and cuboids.

AQA Modular specification reference	Learning objectives	Grade	Resource AQA GCSE Maths Middle sets Student Book; Middle sets Teacher Guide	Common mistakes and misconceptions	Support and homework Middle sets Teacher Guide	Middle sets Practice Book
G1.5, G4.1, G4.3	Calculate the circumference of a circle Calculate the perimeter of compound shapes involving circles or parts of circles	D, C	Section 29.1	Not multiplying by 2 when the radius is given and the diameter is needed.	GPW 29.1	Section 29.1
G1.5, G4.1, G4.3	Calculate the area of a circle Calculate the area of compound shapes involving circles or parts of circles	D, C	Section 29.2	Multiplying by π before squaring.		Section 29.2
G2.4, G4.1, G4.3, G4.4	Calculate the volume of a cylinder Solve problems involving the surface area of cylinders	C	Section 29.3	Multiplying by π before squaring.	GPW 29.3	Section 29.3

G3.4 Convert measurements from one unit to another.

G3.7 Understand and use compound measures.

AQA Modular specification reference	Learning objectives	Grade	Resource AQA GCSE Maths Middle sets Student Book; Middle sets Teacher Guide	Common mistakes and misconceptions	Support and homework	
					Middle sets Teacher Guide	Middle sets Practice Book
G3.4, G3.7	Convert between different units of area Convert between different units of volume	D, C	Section 30.1	Multiplying by 100 when converting from m^3 to cm^3.	GPW 30.1	Section 30.1
G3.7	Calculate average speeds	D	Section 30.2	Not remembering the formulae. Confusing the decimal parts of an hour with hours and minutes (e.g. using 1 hour 45 minutes as 1.45 hours).	GPW 30.2/30.3	Section 30.2
G3.7	*Make calculations using density*	C	*Section 30.3*	*Not remembering the formulae.* *Using the wrong formula.*	*GPW 30.2/30.3*	*Section 30.3*
G3.7	*Recognise formulae for length, area or volume by considering dimensions*	B	*Section 30.4*	*Not appreciating that for an expression to represent a quantity every term must have the same dimension.*		*Section 30.4*

© Pearson Education Limited 2010

Chapter 31 Enlargement and similarity

G1.7 Describe and transform 2D shapes using single or combined rotations, reflections, translations, or enlargements by a positive scale factor and distinguish properties that are preserved under particular transformations.

G1.7h Use positive fractional and negative scale factors.

G1.8 Understand congruence and similarity.

G3.2 Understand the effect of enlargement for perimeter, area and volume of shapes and solids.

AQA Modular specification reference	Learning objectives	Grade	Resource AQA GCSE Maths Middle sets Student Book; Middle sets Teacher Guide	Common mistakes and misconceptions	Support and homework	
					Middle sets Teacher Guide	Middle sets Practice Book
G1.7, G3.2	Enlarge a shape on a grid Enlarge a shape using a centre of enlargement	E, D	Section 31.1	Inaccurately counting squares. Adding the scale factor instead of multiplying by the scale factor. Not using the centre of enlargement.	GPW 31.1	Section 31.1
G1.7h, G3.2	*Enlarge a shape using a fractional scale factor*	*C*	*Section 31.2*	*Not using the centre of enlargement.*	*GPW 31.2*	*Section 31.2*
G1.7h, G1.8	*Understand similarity and the link with enlargement*	*B*	*Section 31.3*	*Incorrectly simplifying the ratio. Not using corresponding sides when making a ratio.*	*GPW 31.3*	*Section 31.3*

Chapter 32 Non-linear graphs

Time: 5 hours [F]; 6 hours [H]

N5.2h Factorise quadratic expressions, including the difference of two squares.

N5.5h Solve quadratic equations.

N6.7h Find the intersection points of the graphs of a linear and quadratic function, knowing that these are the approximate solutions of the corresponding simultaneous equations representing the linear and quadratic functions.

N6.8h Draw, sketch, recognise graphs of simple cubic functions, the reciprocal function $y = 1/x$ with $x \neq 0$, the function $y = k^x$ for integer values of x and simple positive values of k, the circular functions $y = \sin x$ and $y = \cos x$.

N6.11h Construct quadratic and other functions from real life problems and plot their corresponding graphs.

N6.13 Generate points and plot graphs of simple quadratic functions, and use these to find approximate solutions.

AQA Modular specification reference	Learning objectives	Grade	Resource **AQA GCSE Maths Middle sets Student Book; Middle sets Teacher Guide**	Common mistakes and misconceptions	Support and homework **Middle sets Teacher Guide**	**Middle sets Practice Book**
Algebra skills: quadratic expressions (N5.2h, N5.5h)						
N6.11h, N6.13	Draw quadratic graphs Identify the line of symmetry of a quadratic graph Draw and interpret quadratic graphs in real-life contexts	D, C	Section 32.1	Drawing the bottom of the graph flat when a graph has its vertex between two plotted points.	GPW 32.1	Section 32.1
N6.7h	Use a graph to solve quadratic equations	C	Section 32.2	Forgetting to write down all the solutions.		Section 32.2
N6.8h	*Draw cubic graphs* *Use a graph to solve cubic equations*	*B*	*Section 32.3*	*Incorrectly finding the cube of a negative number.* *Forgetting to write down all the solutions.*		*Section 32.3*

© Pearson Education Limited 2010

41

G3.8 Measure and draw lines and angles.

G3.10 Use straight edge and a pair of compasses to do constructions.

G3.11 Construct loci.

AQA Modular specification reference	Learning objectives	Grade	Resource AQA GCSE Maths Middle sets Student Book; Middle sets Teacher Guide	Common mistakes and misconceptions	Support and homework Middle sets Teacher Guide	Middle sets Practice Book
G3.8, G3.10	**Construct perpendiculars** Construct the perpendicular bisector of a line segment **Construct angles of** 90° and **60°** Construct the bisector of an angle	C	Section 33.1	Failing to keep the settings of compasses constant. Rubbing out construction lines. Not using compasses.	GPW 33.1	Section 33.1
G3.11	Construct loci Solve locus problems, including the use of bearings	C	Section 33.2	Confusing a distance from a point with the distance from a line. Making inaccurate constructions. Shading the wrong region.	GPW 33.2	Section 33.2

AQA Modular specification reference	Learning objectives	Grade	Resource AQA GCSE Maths Middle sets Student Book; Middle sets Teacher Guide	Common mistakes and misconceptions	Support and homework	
					Middle sets Teacher Guide	Middle sets Practice Book
G2.1	Understand Pythagoras' theorem	C	Section 34.1	Forgetting that x^2 means $x \times x$, not $x \times 2$.		Section 34.1
G2.1	Calculate the hypotenuse of a right-angled triangle Solve problems using Pythagoras' theorem	C	Section 34.2	Forgetting to take the square root to find the final answer. Not correctly identifying the hypotenuse. Forgetting that Pythagoras' theorem only applies to right-angled triangles.		Section 34.2
G2.1	Calculate the length of a shorter side in a right-angled triangle Solve problems using Pythagoras' theorem	C	Section 34.3	Not correctly identifying the hypotenuse. Forgetting to take the square root to find the final answer. Forgetting that Pythagoras' theorem only applies to right-angled triangles. Not identifying the appropriate information when problems are set in context. Not being able to identify the position of the right angle.		Section 34.3
G2.1	Calculate the length of a line segment *AB*	C	Section 34.4	Subtracting instead of adding the two pairs of coordinates.		Section 34.4

© Pearson Education Limited 2010

Chapter 35 Trigonometry

N1.14h Including trigonometrical functions.

G2.2h Use the trigonometrical ratios and the sine and cosine rules to solve 2D and 3D problems.

G3.6 Understand and use bearings.

AQA Modular specification reference	Learning objectives	Grade	Resource AQA GCSE Maths Middle sets Student Book; Middle sets Teacher Guide	Common mistakes and misconceptions	Support and homework Middle sets Teacher Guide	Middle sets Practice Book
G2.2h, N1.14h	Understand and recall trigonometric ratios in right-angled triangles Know how to enter the trigonometric functions on a calculator	B	Section 35.1	Forgetting that the sine, cosine and tangent ratios only apply to right-angled triangles. Incorrectly using the trigonometric function keys on a calculator.		Section 35.1
G2.2h	Use trigonometric ratios to find lengths in right-angled triangles	B	Section 35.2	Forgetting that the sine, cosine and tangent ratios only apply to right-angled triangles. Not correctly identifying the opposite, adjacent and hypotenuse. Incorrectly using the trigonometric function keys on a calculator.	GPW 35.2	Section 35.2
G2.2h	Use trigonometric ratios to find the angles in right-angled triangles	B	Section 35.3	NOTE: Although sin⁻¹ has been introduced, it is not required at GCSE; therefore, it can be simply said that the inverse is being found. Incorrectly using the trigonometric function keys on a calculator.	GPW 35.3	Section 35.3
G2.2h, G3.6	Use trigonometric ratios and Pythagoras' theorem to solve problems, including the use of bearings	B	Section 35.4	Not identifying the appropriate information when problems are set in context. Drawing a diagram that incorrectly represents the problem. Rounding off values during the intermediate steps of a calculation.		Section 35.4
G2.2h	Solve problems using an angle of elevation or an angle of depression	B	Section 35.5	Not identifying the appropriate information when problems are set in context. Drawing a diagram that incorrectly represents the problem. Rounding off values during the intermediate steps of a calculation.		Section 35.5

G1.5h Know and use circle theorems.

AQA Modular specification reference	Learning objectives	Grade	Resource	Common mistakes and misconceptions	Support and homework	
					Middle sets Teacher Guide	Middle sets Practice Book
			AQA GCSE Maths Middle sets Student Book; Middle sets Teacher Guide			
G1.5h	Use chord and tangent properties to solve problems	B	Section 36.1	Giving answers but not explaining the properties used. Not appreciating that listing the unknown facts can help progress the solution to the problem.		Section 36.1
G1.5h	Use circle theorems to solve geometrical problems	B	Section 36.2	Mistaking chords for diameters and therefore incorrectly identifying the subtended angle as 90°.	GPW 36.2	Section 36.2

End of Modular Scheme of Work

45

© Pearson Education Limited 2010

Longman AQA GCSE Maths Two-year Linear Scheme of Work – OVERVIEW for Middle sets taking Higher Tier

Notes:

1. Y10 Autumn term has 2 hours of revision time available.
2. Y10 Spring term has 3 hours of revision time available.
3. Y10 Summer term has 8 hours of revision time available.
4. The Year 10 scheme assumes an end-of-year exam will be set on the topics covered in Y10.
5. Y11 Autumn term has 4 hours of revision time available.
6. ***Bold Italic*** type shows content that will only be examined at Higher Tier GCSE.
7. Editable Word files are provided on the CD-ROM in the back of this Teacher Guide.

	Chapter	Teaching hours	Grades	AQA Linear specification reference
Y10 AUTUMN TERM	9. Number skills	2 [H]	E, D, C	Working with numbers and the number system: N1.2, N1.4, N1.4h
	13. Decimals	3 [H]	E, D, C, B	Working with numbers and the number system: N1.1, N1.2 Fractions, Decimals and Percentages: N2.3, N2.4
	11. Basic rules of algebra	4 [H]	E, D, C, B	The Language of Algebra: N4.1 Expressions and Equations: N5.1, N5.1h
	21. Angles	3 [H]	E, D, C	Properties of angles and shapes: G1.1, G1.2 Measures and Construction: G3.1, G3.6
	1. Data collection	5 [H]	E, D, C	The Data Handling Cycle: S1 Data Collection: S2.1, S2.2, S2.3, S2.4, S2.5 Data presentation and analysis: S3.1
	22. Measurement 1	3 [H]	E, C	Working with numbers and the number system: N1.3 Measures and Construction: G3.1, G3.3, G3.4
	14. Equations and inequalities	7 [H]	E, D, C, B	Expressions and Equations: N5.4, N5.4h, N5.7, N5.7h
	2. Fractions, decimals and percentages	4 [H]	E, D, C	Working with numbers and the number system: N1.14 Fractions, Decimals and Percentages: N2.6, N2.7
	23. Triangles and constructions	3 [H]	E, D, C	Properties of angles and shapes: G1.1, G1.2, G1.8 Measures and Construction: G3.9, G3.10
	24. Equations, formulae and proof	5 [H]	D, C, B	The Language of Algebra: N4.2 Expressions and Equations: N5.1, N5.4, N5.6, N5.8 Geometrical reasoning and calculation: G2.3, G2.3h
	26. Perimeter, area and volume	4 [H]	E, D, C	Mensuration: G4.1, G4.4

Term	Topic	Lessons	Tier	Specification references
Y10 SPRING TERM	3. Interpreting and representing data	6 [H]	E, C	Data presentation and analysis: S3.2 / Data Interpretation: S4.2, S4.3
	12. Fractions	6 [H]	E, D, C, B	Working with numbers and the number system: N1.2, N1.3, N1.5 / Fractions, Decimals and Percentages: N2.1, N2.2, N2.7
	15. Indices and formulae	6 [H]	E, D, C, B	Working with numbers and the number system: N1.9 / The Language of Algebra: N4.2 / Expressions and Equations: N5.6
	4. Range and averages	4 [H]	E, D, C	Data presentation and analysis: S3.3 / Data Interpretation: S4.1
	17. Sequences and proof	6 [H]	E, D, C	Expressions and Equations: N5.9 / Sequences, Functions and Graphs: N6.1, N6.2
	7. Ratio and proportion	5 [H]	E, D, C, B	Ratio and Proportion: N3.1, N3.2, N3.3
Y10 SUMMER TERM	28. Reflection, translation and rotation	5 [H]	E, D, C	Properties of angles and shapes: G1.6, G1.7 / Vectors: G5.1
	20. Number skills revisited	2 [H]		Working with numbers and the number system: N1.3, N1.4, N1.14 / Fractions, decimals and Percentages: N2.1, N2.5, N2.7 / Ratio and Proportion: N3.1
	16. Percentages	5 [H]	E, D, C, B	Fractions, Decimals and Percentages: N2.5, N2.7, N2.7h
	25. Quadrilaterals and other polygons	4 [H]	E, D, C	Expressions and Equations: N5.4 / Sequences, Functions and Graphs: N6.3 / Properties of angles and shapes: G1.2, G1.3, G1.4
	5. Probability	7 [H]	E, D, C, B	Data presentation and analysis: S3.1 / Probability: S5.1, S5.2, S5.3, S5.4, S5.5h, S5.6h, S5.7, S5.8, S5.9
	8. Complex calculations	5 [H]	C, B	Working with numbers and the number system: N1.10h / Fractions, Decimals and Percentages: N2.7h
Y11 AUTUMN TERM	18. Linear graphs	7 [H]	E, D, C, B	Expressions and Equations: N5.4h, N5.7h / Sequences, Functions and Graphs: N6.3, N6.4, N6.5h, N6.6h, N6.11, N6.12
	27. 3-D objects	1 [H]	E, D	Geometrical reasoning and calculation: G2.4
	30. Measurement 2	4 [H]	D, C, B	Measures and Construction: G3.4, G3.7
	10. Factors, powers and standard form	6 [H]	E, D, C, B	Working with numbers and the number system: N1.6, N1.7, N1.8, N1.9, N1.9h, N1.10h
	6. Cumulative frequency	6 [H]	B	Data presentation and analysis: S3.2h, S3.3h / Data Interpretation: S4.4
	29. Circles and cylinders	6 [H]	D, C	Properties of angles and shapes: G1.5 / Geometrical reasoning and calculation: G2.4 / Mensuration: G4.1, G4.3, G4.4
	31. Enlargement and similarity	5 [H]	E, D, C, B	Properties of angles and shapes: G1.7, G1.7h, G1.8 / Measures and Construction: G3.2
	19. Quadratic equations	6 [H]	B	Expressions and Equations: N5.2h, N5.5h

47

© Pearson Education Limited 2010

	Topic	Hours	Grade	Specification references
Y11 SPRING TERM	32. Non-linear graphs	6 [H]	D, C, B	Expressions and Equations: N5.2h, N5.5h Sequences, Functions and Graphs: N6.7h, N6.8h, N6.11h, N6.13
	34. Pythagoras' theorem	4 [H]	C	Geometrical reasoning and calculation: G2.1
	33. Constructions and loci	4 [H]	C	Measures and Construction: G3.8, G3.10, G3.11
	35. Trigonometry	7 [H]	B	Working with numbers and the number system: N1.14h Geometrical reasoning and calculation: G2.2h Measures and Construction: G3.6
	36. Circle theorems	4 [H]	B	Properties of angles and shapes: G1.5h
Y11 SUMMER TERM	REVISION FOR JUNE EXAMS (29 HOURS)			

Longman AQA GCSE Maths Two-year Linear Scheme of Work – OVERVIEW for Middle sets taking Foundation Tier

Notes:

1. Y10 Autumn term has 4 hours of revision time available.
2. Y10 Spring term has 4 hours of revision time available.
3. Y10 Summer term has 4 hours of revision time available.
4. The Year 10 scheme assumes an end-of-year exam will be set on the topics covered in Y10.
5. Y11 Autumn term has 5 hours of revision time available.
6. ***Bold Italic*** type shows content that will only be examined at Higher Tier GCSE.
7. Editable Word files are provided on the CD-ROM in the back of this Teacher Guide.

	Chapter	Teaching hours	Grades	AQA Linear specification reference
Y10 AUTUMN TERM	9. Number skills	4 [F]	E, D, C	Working with numbers and the number system: N1.2, N1.4, N1.4h
	13. Decimals	3 [F]	E, D, C, B	Working with numbers and the number system: N1.1, N1.2 Fractions, Decimals and Percentages: N2.3, N2.4
	11. Basic rules of algebra	6 [F]	E, D, C, B	The Language of Algebra: N4.1 Expressions and Equations: N5.1, N5.1h
	21. Angles	5 [F]	E, D, C	Properties of angles and shapes: G1.1, G1.2 Measures and Construction: G3.1, G3.6
	1. Data collection	6 [F]	E, D, C	The Data Handling Cycle: S1 Data Collection: S2.1, S2.2, S2.3, S2.4, S2.5 Data presentation and analysis: S3.1
	22. Measurement 1	4 [F]	E, C	Working with numbers and the number system: N1.3 Measures and Construction: G3.1, G3.3, G3.4
	14. Equations and inequalities	7 [F]	E, D, C, B	Expressions and Equations: N5.4, N5.4h, N5.7, N5.7h
	2. Fractions, decimals and percentages	6 [F]	E, D, C	Working with numbers and the number system: N1.14 Fractions, Decimals and Percentages: N2.6, N2.7
Y10 SPRING TERM	23. Triangles and constructions	4 [F]	E, D, C	Properties of angles and shapes: G1.1, G1.2, G1.8 Measures and Construction: G3.9, G3.10
	24. Equations, formulae and proof	6 [F]	D, C, B	The Language of Algebra: N4.2 Expressions and Equations: N5.1, N5.4, N5.6, N5.8 Geometrical reasoning and calculation: G2.3, G2.3h
	26. Perimeter, area and volume	6 [F]	E, D, C	Mensuration: G4.1, G4.4
	3. Interpreting and representing data	8 [F]	E, D, C	Data presentation and analysis: S3.2 Data Interpretation: S4.2, S4.3
	12. Fractions	8 [F]	E, D, C, B	Working with numbers and the number system: N1.2, N1.3, N1.5 Fractions, Decimals and Percentages: N2.1, N2.2, N2.7

© Pearson Education Limited 2010

Term	Topic	Lessons	Levels	Specification references
Y10 SUMMER TERM	15. Indices and formulae	6 [F]	E, D, C, B	Working with numbers and the number system: N1.9 The Language of Algebra: N4.2 Expressions and Equations: N5.6
	4. Range and averages	6 [F]	E, D, C	Data presentation and analysis: S3.3 Data Interpretation: S4.1
	17. Sequences and proof	8 [F]	E, D, C	Expressions and Equations: N5.9 Sequences, Functions and Graphs: N6.1, N6.2
	7. Ratio and proportion	6 [F]	E, D, C, B	Ratio and Proportion: N3.1, N3.2, N3.3
	28. Reflection, translation and rotation	6 [F]	E, D, C	Properties of angles and shapes: G1.6, G1.7 Vectors: G5.1
Y11 AUTUMN TERM	20. Number skills revisited	3 [F]		Working with numbers and the number system: N1.3, N1.4, N1.14 Fractions, decimals and Percentages: N2.1, N2.5, N2.7 Ratio and Proportion: N3.1
	16. Percentages	6 [F]	E, D, C, B	Fractions, Decimals and Percentages: N2.5, N2.7, N2.7h
	25. Quadrilaterals and other polygons	6 [F]	E, D, C	Expressions and Equations: N5.4 Sequences, Functions and Graphs: N6.3 Properties of angles and shapes: G1.2, G1.3, G1.4
	5. Probability	8 [F]	E, D, C, B	Data presentation and analysis: S3.1 Probability: S5.1, S5.2, S5.3, S5.4, S5.5h, S5.6h, S5.7, S5.8, S5.9
	8. Complex calculations	2 [F]	C, B	Working with numbers and the number system: N1.10h Fractions, Decimals and Percentages: N2.7h
	18. Linear graphs	7 [F]	E, D, C, B	Expressions and Equations: N5.4h, N5.7h Sequences, Functions and Graphs: N6.3, N6.4, N6.5h, N6.6h, N6.11, N6.12
	27. 3-D objects	2 [F]	E, D	Geometrical reasoning and calculation: G2.4
	30. Measurement 2	3 [F]	D, C, B	Measures and Construction: G3.4, G3.7
	10. Factors, powers and standard form	3 [F]	E, D, C, B	Working with numbers and the number system: N1.6, N1.7, N1.8, N1.9, N1.9h, N1.10h

6. Cumulative frequency	0 [F]	B	Data presentation and analysis: S3.2h, S3.3h / Data Interpretation: S4.4
29. Circles and cylinders	7 [F]	D, C	Properties of angles and shapes: G1.5 / Geometrical reasoning and calculation: G2.4 / Mensuration: G4.1, G4.3, G4.4
31. Enlargement and similarity	2 [F]	E, D, C, B	Properties of angles and shapes: G1.7, G1.7h, G1.8 / Measures and Construction: G3.2
19. Quadratic equations	0 [F]	B	Expressions and Equations: N5.2h, N5.5h
32. Non-linear graphs	5 [F]	D, C, B	Expressions and Equations: N5.2h, N5.5h / Sequences, Functions and Graphs: N6.7h, N6.8h, N6.11h, N6.13
34. Pythagoras' theorem	6 [F]	C	Geometrical reasoning and calculation: G2.1
33. Constructions and loci	5 [F]	C	Measures and Construction: G3.8, G3.10, G3.11
35. Trigonometry	0 [F]	B	Working with numbers and the number system: N1.14h / Geometrical reasoning and calculation: G2.2h / Measures and Construction: G3.6
36. Circle theorems	0 [F]	B	Properties of angles and shapes: G1.5h

Y11 SPRING TERM

REVISION FOR JUNE EXAMS (29 HOURS)

Y11 SUMMER TERM

Detailed Linear Scheme of Work begins on page 52

© Pearson Education Limited 2010

Longman AQA GCSE Maths Two-year Linear Scheme of Work – For Middle sets taking Higher or Foundation Tier

Bold italic text indicates content that will only be examined at Higher Tier

Chapter 9 Number skills

Time: 4 hours [F]; 2 hours [H]

N1.2 Add, subtract, multiply and divide any number.

N1.4 Approximate to a given power of 10, up to three decimal places and one significant figure.

N1.4h Approximate to specified or appropriate degrees of accuracy, including a given number of decimal places and significant figures.

AQA Linear specification reference	Learning objectives	Grade	Resource	Common mistakes and misconceptions	Support and homework	
			AQA GCSE Maths Middle sets Student Book; Middle sets Teacher Guide		Middle sets Teacher Guide	Middle sets Practice Book
Number skills: adding and subtracting (N1.2); multiplying and dividing (N1.2)						
N1.2, N1.4	Multiply whole numbers using written methods. Use repeated subtraction for division of whole numbers. Round up or down in context	E, D	Section 9.1	Forgetting to add the numbers to find the final answer when using the grid method. Forgetting the zero when multiplying by tens when using the standard method. Writing 3.6 to represent 3 remainder 6. Not giving an answer in the context of the problem.	GPW 9.1	Section 9.1
N1.4, N1.4h	Check and estimate answers to problems. Estimate answers to problems involving decimals. Make estimates and approximations of calculations	E, D, C	Section 9.2	Finding an approximate value independent of the context in which it is set. Working out the actual answer instead of an approximation.		Section 9.2
N1.2	Multiply and divide negative numbers	E	Section 9.3	Applying the 'general rules' (*the signs are different, so the answer is negative; the signs are the same, so the answer is positive*) without constraint. Applying the rules for multiplying/dividing negative numbers to adding/subtracting negative numbers.		Section 9.3

N1.1 Understand integers and place value to deal with arbitrarily large positive numbers.

N1.2 Add, subtract, multiply and divide any number.

N2.3 Use decimal notation and recognise that each terminating decimal is a fraction.

N2.4 Recognise that recurring decimals are exact fractions, and that some exact fractions are recurring decimals.

AQA Linear specification reference	Learning objectives	Grade	Resource AQA GCSE Maths Middle sets Student Book; Middle sets Teacher Guide	Common mistakes and misconceptions	Support and homework	
					Middle sets Teacher Guide	Middle sets Practice Book
N1.1, N1.2	Add and subtract decimal numbers	E	Section 13.1	Not lining up the decimal points. Not recording the 'carry over' and forgetting to add it on. Not reducing a number during an exchange.	GPW 13.1	Section 13.1
N2.3	Convert decimals to fractions	D	Section 13.2	Working with the incorrect power of 10. Not giving answers in the simplest form.	GPW 13.2	Section 13.2
N1.2	Multiply and divide decimal numbers	D, C	Section 13.3	Working out the equivalent whole-number multiplication but forgetting to return to the decimal calculation at the end. Confusing multiplication with the rules for addition, writing a long multiplication with decimal points underneath each other.	GPW 13.3	Section 13.3
N2.3, N2.4	*Convert fractions to decimals* *Recognise recurring decimals* *Understand how recurring decimals relate to fractions*	*D, C, B*	*Section 13.4*	*Confusing 0.3 with $\frac{1}{3}$.* *Not understanding that recurring decimals are a form of exact maths and therefore rounding answers.*	*GPW 13.4*	*Section 13.4*

© Pearson Education Limited 2010

N4.1 Distinguish the different roles played by letter symbols in algebra, using the correct notation.

N5.1 Manipulate algebraic expressions by collecting like terms, by multiplying a single term over a bracket, and by taking out common factors.

N5.1h Multiply two linear expressions.

AQA Linear specification reference	Learning objectives	Grade	Resource AQA GCSE Maths Middle sets Student Book; Middle sets Teacher Guide	Common mistakes and misconceptions	Support and homework Middle sets Teacher Guide	Middle sets Practice Book
Algebra skills: writing and simplifying expressions (N4.1, N5.1)						
N4.1, N5.1	Simplify algebraic expressions by collecting like terms	E	Section 11.1	Failing to comprehend that $x = 1x$. Combining unlike terms.		Section 11.1
N5.1	Multiply together two simple algebraic expressions	E	Section 11.2	Treating terms in m^2 and in m as like terms (e.g. simplifying $3m^2 + m$ wrongly to $4m^2$).		Section 11.2
N5.1	Multiply terms in a bracket by a number outside the bracket Multiply terms in a bracket by a term that includes a letter	D	Section 11.3	Forgetting to multiply the second term in the bracket by the term outside (e.g. expanding $2(x + 3)$ as $2x + 3$), or ignoring minus signs (e.g. writing $3(m - 2)$ as $3m + 6$).	GPW 11.3	Section 11.3
N5.1	Simplify expressions involving brackets	D, C	Section 11.4	Forgetting to multiply the second term in the bracket by the term outside. Getting the wrong signs when multiplying negative values.		Section 11.4
N5.1	Recognise factors of algebraic terms Simplify algebraic expressions by taking out common factors	D	Section 11.5	Not realising that x is a factor of x and x^2. Not taking out the highest common factor.	GPW 11.5	Section 11.5
N5.1h	***Multiply together two algebraic expressions with brackets*** ***Square a linear expression***	***C, B***	***Section 11.6***	***Forgetting to multiply pairs of terms.***	***GPW 11.6***	***Section 11.6***

G1.1 Recall and use properties of angles at a point, angles at a point on a straight line (including right angles), perpendicular lines, and opposite angles at a vertex.

G1.2 Understand and use the angle properties of parallel and intersecting lines, triangles and quadrilaterals.

G3.1 Use and interpret maps and scale drawings.

G3.6 Understand and use bearings.

AQA Linear specification reference	Learning objectives	Grade	Resource AQA GCSE Maths Middle sets Student Book; Middle sets Teacher Guide	Common mistakes and misconceptions	Support and homework	
					Middle sets Teacher Guide	Middle sets Practice Book
Geometry skills: angles (G1.1)						
G1.1	Calculate angles around a point Recognise vertically opposite angles	E	Section 21.1	Measuring rather than calculating angles.		Section 21.1
G1.2	Recognise corresponding and alternate angles Calculate angles in diagrams with parallel lines	D	Section 21.2	Confusing alternate and corresponding angles.	GPW 21.2	Section 21.2
G3.1, G3.6	Use three-figure bearing notation Measure the bearing from one place to another Plot a bearing Calculate bearings for return journeys Draw and interpret scale diagrams to represent journeys	E, D, C	Section 21.3	Confusing which angles need to be found. Not realising that some of the angles asked for can simply be read off the diagram.	GPW 21.3	Section 21.3

© Pearson Education Limited 2010

Chapter 1 Data collection

Time: 6 hours [F]; 5 hours [H]

S1 Understand and use the statistical problem solving process which involves

- specifying the problem and planning
- processing and presenting the data
- collecting data
- interpreting and discussing the results.

S2.1 Types of data: qualitative, discrete, continuous. Use of grouped and ungrouped data.

S2.2 Identify possible sources of bias.

S2.3 Design an experiment or survey.

S2.4 Design data collection sheets distinguishing between different types of data.

S2.5 Extract data from printed tables and lists.

S3.1 Design and use two-way tables for grouped and ungrouped data.

AQA Linear specification reference	Learning objectives	Grade	Resource AQA GCSE Maths Middle sets Student Book; Middle sets Teacher Guide	Common mistakes and misconceptions	Support and homework	
					Middle sets Teacher Guide	Middle sets Practice Book
S1	Learn about the data handling cycle Know how to write a hypothesis	D	Section 1.1	Formulating a hypothesis that cannot be tested. Thinking that a hypothesis is not valuable if it is eventually proved false.		Section 1.1
S2.3, S2.4	Know where to look for information	D	Section 1.2	Not realising that data collected by a third party (even if the results of a survey or experiment) is classed as secondary data.		Section 1.2
S2.1	Be able to identify different types of data	D	Section 1.3	Not appreciating that some data can be treated as either discrete or continuous depending on the context (e.g. age – this is really continuous, but is often treated as discrete, such as when buying child or adult tickets).		Section 1.3
S2.4	Work out methods for gathering data efficiently	E	Section 1.4	Using shortcuts in the tallying process – counting up the items in each class, rather than tallying items one by one.		Section 1.4
S2.4, S2.5	Work out methods for gathering data that can take a wide range of values	D	Section 1.5	Using overlapping class intervals. Recording data which is on the boundary of a class interval in the wrong class.		Section 1.5
S2.5, S3.1	Work out methods for recording related data	D	Section 1.6	Not checking that the totals in two-way tables add up.		Section 1.6
S2.3, S2.4	Learn how to write good questions to find out information	C	Section 1.7	Using overlapping classes, or gaps between classes, for response options.		Section 1.7
S2.2, S2.3, S2.4	Know the techniques to use to get a reliable sample	C	Section 1.8	Mistaking biased samples for random samples.		Section 1.8

Chapter 22 Measurement 1

Time: 4 hours [F]; 3 hours [H]

N1.3 Understand and use number operations and the relationships between them, including inverse operations and hierarchy of operations.

G3.1 Use and interpret maps and scale drawings.

G3.3 Interpret scales on a range of measuring instruments and recognise the inaccuracy of measurements.

G3.4 Convert measurements from one unit to another.

AQA Linear specification reference	Learning objectives	Grade	Resource **AQA GCSE Maths Middle sets Student Book; Middle sets Teacher Guide**	Common mistakes and misconceptions	Support and homework	
					Middle sets Teacher Guide	**Middle sets Practice Book**
N1.3	Solve problems involving times, dates and timetables	E	Section 22.1	Confusing the decimal parts of an hour with hours and minutes (e.g. writing 1.25 hours as 1 hour 25 minutes) and vice versa.	GPW 22.1	Section 22.1
G3.4	Know and use approximate metric equivalents of pounds, feet, miles, pints and gallons	E	Section 22.2	Not considering the relative size of units when deciding whether to multiply or divide.	GPW 22.2	Section 22.2
G3.1	Use and interpret maps and scale drawings	E	Section 22.3	Missing out steps when converting between (for example) km and cm. Not making allowances when measurements are given in a variety of units.		Section 22.3
G3.3	Recognise that measurements given to the nearest whole unit may be inaccurate by up to one half unit in either direction	C	Section 22.4	Difficulty comprehending the definition of the upper bound, since, for example, 146.5 rounds to 147.		Section 22.4

© Pearson Education Limited 2010

Chapter 14 Equations and inequalities

N5.4 Set up and solve simple linear equations.

N5.4h Including simultaneous equations in two unknowns.

N5.7 Solve linear inequalities in one variable and represent the solution set on a number line.

N5.7h Solve linear inequalities in two variables, and represent the solution set on a suitable diagram.

AQA Linear specification reference	Learning objectives	Grade	Resource AQA GCSE Maths Middle sets Student Book; Middle sets Teacher Guide	Common mistakes and misconceptions	Support and homework Middle sets Teacher Guide	Middle sets Practice Book
N5.4	Solve two-step equations like $2x - 1 = 11$	E, D	Section 14.1	Not appreciating that an equation can be written in different but equivalent formats (e.g. $2a + 7 = 9 \to 7 + 2a = 9 \to 9 = 2a + 7$).	GPW 14.1a–14.4a GPW 14.1b–14.4b	Section 14.1
N5.4	Write and solve equations	E, D	Section 14.2	Not following a question carefully when writing an equation to represent a problem.	GPW 14.1a–14.4a GPW 14.1b–14.4b	Section 14.2
N5.4	Solve equations involving brackets	D, C	Section 14.3	Forgetting to multiply the second term in the bracket by the term outside. Getting the wrong signs when multiplying negative numbers. Incorrectly simplifying after expanding the bracket.	GPW 14.1a–14.4a GPW 14.1b–14.4b	Section 14.3
N5.4	Solve equations with an unknown on both sides	D, C	Section 14.4	Introducing errors when there are a negative number of unknowns on either side of the equation.	GPW 14.1a–14.4a GPW 14.1b–14.4b	Section 14.4
N5.4	*Solve equations involving fractions*	*C, B*	*Section 14.5*	*Incorrectly cancelling after multiplying by the LCM.* *Solving out of order (e.g. $\frac{x+2}{4} = 8$: trying to do −2 first).*		*Section 14.5*

N5.7, N5.7h	Represent inequalities on a number line Write down whole-number values for unknowns in an inequality Solve inequalities	E, D, C, B	Section 14.6	Not reversing the sign when multiplying or dividing by a negative. Confusing the convention of an open circle for a strict inequality and a closed circle for an included boundary.	Section 14.6
N5.4h	Solve a pair of simultaneous equations	B	Section 14.7	Adding equations when they should be subtracted, and vice versa.	Section 14.7

© Pearson Education Limited 2010

Chapter 2 Fractions, decimals and percentages

Time: 6 hours [F]; 4 hours [H]

N1.14 Use calculators effectively and efficiently, including statistical functions.

N2.6 Interpret fractions, decimals and percentages as operators.

N2.7 Calculate with fractions, decimals and percentages.

AQA Linear specification reference	Learning objectives	Grade	Resource AQA GCSE Maths Middle sets Student Book; Middle sets Teacher Guide	Common mistakes and misconceptions	Support and homework Middle sets Teacher Guide	Middle sets Practice Book
N2.7	Find a fraction of an amount with a calculator / Find a fraction of an amount with a calculator in more complex situations	E, D	Section 2.1	Incorrectly inputting numbers on the calculator. Being unsure of what to work out when a fraction calculation is set in context.	GPW 2.1	Section 2.1
N2.7	Write one quantity as a fraction of another	D	Section 2.2	Not making the denominator the total in questions involving a number of quantities. Working with quantities in different units. Incorrectly cancelling down.	GPW 2.2	Section 2.2
N1.14	Use the fraction key on a calculator / Use the fraction key on a calculator with mixed numbers	E, D	Section 2.3	Not recognising or know how to use the fraction key on a calculator. Misinterpreting a mixed number on a calculator display.		Section 2.3
N2.7	Find a percentage of an amount with a calculator / Find percentages of amounts in more complex situations	E, D	Section 2.4	Thinking that percentages over 100% cannot exist. Treating a percentage such as 0.05% as though it were 5%. Adding the percentage to the cost when finding a percentage increase (e.g. £315 + 15% VAT = £330).		Section 2.4
N2.7	Write one quantity as a percentage of another / Write one quantity as a percentage of another in more complex situations	D, C	Section 2.5	Not using the original amount as the denominator, when finding a percentage difference. Working with quantities in different units.		Section 2.5
N2.6, N2.7	Calculate a percentage increase or decrease	D	Section 2.6	Giving the actual increase/decrease as the answer when the amount after the increase/decrease is what is required. Using the multiplier as 1.5 rather than 1.05 for an increase of 5%.		Section 2.6
N2.7	Understand and use a retail prices index / Understand and use a retail prices index in more complex situations	D, C	Section 2.7	Using a previously found price instead of the base year price.	GPW 2.7	Section 2.7

Chapter 23 Triangles and constructions

Time: 4 hours [F]; 3 hours [H]

G1.1 Recall and use properties of angles at a point, angles at a point on a straight line (including right angles), perpendicular lines, and opposite angles at a vertex.

G1.2 Understand and use the angle properties of parallel and intersecting lines, triangles and quadrilaterals.

G1.8 Understand congruence and similarity.

G3.9 Draw triangles and other 2D shapes using a ruler and protractor.

G3.10 Use straight edge and a pair of compasses to do constructions.

AQA Linear specification reference	Learning objectives	Grade	Resource AQA GCSE Maths Middle sets Student Book; Middle sets Teacher Guide	Common mistakes and misconceptions	Support and homework Middle sets Teacher Guide	Middle sets Practice Book
G1.1, G1.2	Solve angle problems in triangles Solve angle problems in triangles involving algebra	E, D	Section 23.1	Not realising when a triangle is isosceles and thinking that the problem cannot be solved. Trying to do too many steps in one go when answering algebra-based questions.	GPW 23.1	Section 23.1
G3.9, G3.10	Draw triangles accurately when given the length of all three sides Draw triangles accurately when at least one angle is given	E, D	Section 23.2	Inaccurately using a protractor or compasses. Not completing the triangle by drawing the third side. Rubbing out construction lines.		Section 23.2
G1.8	Recognise and explain how triangles are congruent	C	Section 23.3	Thinking that two triangles are congruent when they are not (due to the relative positions of side lengths or angles being in different positions). Incorrectly assuming that by showing that the three sets of angles are the same in both triangles congruency is proved.		Section 23.3

© Pearson Education Limited 2010

Chapter 24 Equations, formulae and proof

N4.2 Distinguish in meaning between the words 'equation', 'formula', and 'expression'.

N5.1 Manipulate algebraic expressions by collecting like terms, by multiplying a single term over a bracket, and by taking out common factors.

N5.4 Set up and solve simple linear equations.

N5.6 Derive a formula, substitute numbers into a formula.

N5.8 Use systematic trail and improvement to find approximate solutions of equations where there is no simple analytical method of solving them.

G2.3 Justify simple geometrical properties.

G2.3h Simple geometrical proofs.

AQA Linear specification reference	Learning objectives	Grade	Resource **AQA GCSE Maths Middle sets Student Book; Middle sets Teacher Guide**	Common mistakes and misconceptions	Support and homework	
					Middle sets Teacher Guide	**Middle sets Practice Book**
Algebra skills: expressions (N4.2, N5.1); brackets (N5.1); solving equations (N5.4); formulae (N5.6)						
N5.4, N5.6	Write your own formulae and equations Substitute into a formula to solve problems Set up and solve equations Change the subject of a formula	D, C, B	Section 24.1	Failing to consider the different terms of an expression when changing the subject of a formula (e.g. $W = \frac{1}{2}x + 3 \Rightarrow 2W = x + 3$). Not using brackets or a clear division (e.g. rewriting $c = 2a + 5$ as $a = c - 5 \div 2$). Not using the inverse operation (e.g. $x + y = z$ becomes $x = z + y$).	GPW 24.1	Section 24.1
G2.3, G2.3h	Prove simple results from geometry	C	Section 24.2	Not laying out answers in an organised way. Not providing reasons for each stage of the working.		Section 24.2
N5.8	Use trial and improvement to find solutions to equations	C	Section 24.3	Not checking the mid-point to determine which of two values is correct (e.g. choosing between $x = 3.3$ and $x = 3.4$ based on the value of the function and the desired output). Using the value of the equation as the answer rather than the value of the variable.	GPW 24.3	Section 24.3

G4.1 Calculate perimeters and areas of shapes made from triangles and rectangles.

G4.4 Calculate volumes of right prisms and of shapes made from cubes and cuboids.

AQA Linear specification reference	Learning objectives	Grade	Resource AQA GCSE Maths Middle sets Student Book; Middle sets Teacher Guide	Common mistakes and misconceptions	Support and homework	
					Middle sets Teacher Guide	Middle sets Practice Book
G4.1	Find the perimeter and area of rectangles, parallelograms, triangles and trapezia	E, D	Section 26.1	Not making rough estimates of areas as a check to avoid arithmetical errors. Incorrectly converting between units. Using measurements in different units.	GPW 26.1	Section 26.1
G4.1	Find the perimeter and area of compound shapes	E, D	Section 26.2	Incorrectly calculating missing lengths. Adding areas instead of subtracting.		Section 26.2
G4.1, G4.4	Find the volume and surface area of a prism	E, D, C	Section 26.3	Confusing volume and surface area.	GPW 26.3	Section 26.3

© Pearson Education Limited 2010

Chapter 3 Interpreting and representing data
Time: 8 hours [F]; 6 hours [H]

S3.2 Produce charts and diagrams for various data types. Scatter graphs, stem-and-leaf, tally charts, pictograms, bar charts, dual bar charts, pie charts, line graphs, frequency polygons, histograms with equal class intervals.

S4.2 Look at data to find patterns and exceptions.

S4.3 Recognise correlation and draw and/or use lines of best fit by eye, understanding what they represent.

AQA Linear specification reference	Learning objectives	Grade	Resource AQA GCSE Maths Middle sets Student Book; Middle sets Teacher Guide	Common mistakes and misconceptions	Support and homework Middle sets Teacher Guide	Middle sets Practice Book
S3.2	Draw a pie chart	E	Section 3.1	Not drawing the angles in the pie chart accurately or using the appropriate scale on the protractor. Measuring each angle from the same starting point.	GPW 3.1	Section 3.1
S3.2	Draw stem-and-leaf diagrams	D	Section 3.2	Forgetting to put a key and order the leaves. Forgetting to recombine the stem and leaf and just giving the leaf as the value.		Section 3.2
S3.2, S4.2	Draw a scatter diagram on a given grid. Interpret points on a scatter diagram	D	Section 3.3	Assuming that all the plotted points must be joined with a line. Drawing the diagram without spending time working out the best scale.		Section 3.3
S4.3	Draw a line of best fit on a scatter diagram. Describe types of correlation. Use the line of best fit	D, C	Section 3.4	Trying to make the line of best fit go through the origin, rather than drawing it appropriately. Not appreciating correlation in terms of 'positive' and 'negative'.		Section 3.4
S3.2	Draw frequency diagrams for grouped data	D	Section 3.5	Using grouped labels on the data axes (e.g. 15–20, rather than the ends of the bar being clearly marked with a 15 at one end and a 20 at the other end).		Section 3.5
S3.2	Draw frequency polygons for grouped data	C	Section 3.6	Using a grouped label on the horizontal axis rather than a continuous scale. Plotting the upper bound instead of the mid-point.	GPW 3.6	Section 3.6

Chapter 12 Fractions

Time: 8 hours [F]; 6 hours [H]

N1.2 Add, subtract, multiply and divide any number.

N1.3 Understand and use number operations and the relationships between them, including inverse operations and hierarchy of operations.

N1.5 Order rational numbers.

N2.1 Understand equivalent fractions, simplifying a fraction by cancelling all common factors.

N2.2 Add and subtract fractions.

N2.7 Calculate with fractions, decimals and percentages.

AQA Linear specification reference	Learning objectives	Grade	Resource **AQA GCSE Maths Middle sets Student Book; Middle sets Teacher Guide**	Common mistakes and misconceptions	Support and homework **Middle sets Teacher Guide**	**Middle sets Practice Book**
N1.5, N2.1	Compare fractions with different denominators	E, D	Section 12.1	Multiplying the denominator but not the numerator when finding equivalent fractions.	GPW 12.1	Section 12.1
N2.2	Add and subtract fractions when one denominator is a multiple of the other Add and subtract fractions when both denominators have to be changed	E, D	Section 12.2	Adding/subtracting the denominators as well as the numerators. Not converting to equivalent fractions to make the denominators the same.		Section 12.2
N2.2	Add and subtract mixed numbers	C	Section 12.3	Incorrectly converting a mixed number to an improper fraction. Not converting the final answer back to a mixed number.	GPW 12.3	Section 12.3
N1.2	Multiply a fraction by a fraction	E	Section 12.4	Multiplying diagonally as though 'cross-multiplying' is being done (e.g. $\frac{2}{3} \times \frac{5}{6} = \frac{12}{15}$).		Section 12.4
N2.7	Multiply a whole number by a mixed number Multiply a fraction by a mixed number Multiply a mixed number by a mixed number	D, C, B	Section 12.5	Multiplying both the numerator and the denominator by the whole number (e.g. $3 \times \frac{5}{6} = \frac{15}{18}$).		Section 12.5
N1.3	Find the reciprocal of a whole number, a decimal or a fraction	C	Section 12.6	Leaving denominators as decimal numbers. Not simplifying answers when asked to do so.	GPW 12.6	Section 12.6
N1.2	Divide a whole number or a fraction by a fraction Divide mixed numbers or fractions by whole numbers Divide mixed numbers by mixed numbers	D, C, B	Section 12.7	Finding the reciprocal of the wrong fraction, or finding the reciprocal of both fractions.		Section 12.7

© Pearson Education Limited 2010

Chapter 15 Indices and formulae

N1.9 Index laws for multiplication and division of integer powers.

N4.2 Distinguish in meaning between the words 'equation', 'formula', and 'expression'.

N5.6 Derive a formula, substitute numbers into a formula and change the subject of a formula.

AQA Linear specification reference	Learning objectives	Grade	Resource AQA GCSE Maths Middle sets Student Book; Middle sets Teacher Guide	Common mistakes and misconceptions	Support and homework Middle sets Teacher Guide	Middle sets Practice Book
N1.9	Use index notation in algebra Use index notation when multiplying or dividing algebraic terms	E, D, C	Section 15.1	Not realising that x means x^1, or that a number divided by 1 equals the number (e.g. $6 \div 1 = 6$).		Section 15.1
N1.9	Use index laws to multiply and divide powers in algebra Raise a number or variable to the power of 1 or 0 Use index laws for raising a power to another power	C, B	Section 15.2	Confusing unit and zero powers (e.g. stating that x means x^0, or that $x^0 = 0$).	GPW 15.2	Section 15.2
N4.2	Use algebra to write formulae in different situations	E, D	Section 15.3	Not seeing the 'general' case.		Section 15.3
N4.2, N5.6	Substitute numbers to work out the value of simple algebraic expressions Substitute numbers into expressions involving brackets and powers	E, D, C	Section 15.4	Incorrectly substituting values into expressions (e.g. substituting $a = 6$ into the expression $4a$, writing 46 and assuming it is forty-six). Ignoring BIDMAS.		Section 15.4
N4.2, N5.6	Substitute numbers into a variety of formulae	E, D	Section 15.5	Not realising that $\frac{n}{10}$ means $n \div 10$, or that $\frac{1}{2} \times 6$ means $\frac{1}{2}$ of $6 = 3$.	GPW 15.5	Section 15.5
N5.6	*Rearrange a formula to make a different variable the subject of the formula*	*C, B*	*Section 15.6*	*Not using brackets or a clear division (e.g. rewriting $c = 2a + 5$ as $a = c - 5 \div 2$).* *Not using the inverse operation (e.g. $x + y = z$ becomes $x = z + y$).*	*GPW 15.6*	*Section 15.6*

Chapter 4 Range and averages

S3.3 Calculate median, mean, range, mode and modal class.

S4.1 Interpret a wide range of graphs and diagrams and draw conclusions.

AQA Linear specification reference	Learning objectives	Grade	Resource AQA GCSE Maths Middle sets Student Book; Middle sets Teacher Guide	Common mistakes and misconceptions	Support and homework	
					Middle sets Teacher Guide	Middle sets Practice Book
S3.3, S4.1	Find the mean, median and range from a set of data, including data given in a stem and leaf diagram	E, D	Section 4.1	Omitting units when writing averages or range. Forgetting to include the stem when reading the median from a stem and leaf diagram.	GPW 4.1	Section 4.1
S3.3, S4.1	Calculate the mode, median and range from an ungrouped frequency table	E	Section 4.2	Confusing the frequencies and the data values.		Section 4.2
S3.3, S4.1	Calculate the mean from an ungrouped frequency table	D, C	Section 4.3	Dividing by the number of rows in the frequency table (i.e. the number of different data values), not by the sum of the frequencies.		Section 4.3
S3.3, S4.1	Find the modal class from a grouped frequency table. Estimate the range from a grouped frequency table. Work out the class interval which contains the median from data given in a grouped frequency table	D, C	Section 4.4	Not appreciating that the statistics calculated from grouped frequency tables are estimates. Not understanding that the estimate for the range is an upper limit.	GPW 4.4/4.5	Section 4.4
S3.3, S4.1	Estimate the mean of data given in a grouped frequency table	C	Section 4.5	Incorrectly calculating the mid-points of class intervals for grouped discrete data (e.g. the mid-point of the class interval 10–19 is 14.5, not 15). Interpreting 'find an estimate for the mean' as 'guess the mean'.	GPW 4.4/4.5	Section 4.5

© Pearson Education Limited 2010

Chapter 17 Sequences and proof

Time: 8 hours [F]; 6 hours [H]

N5.9 Use algebra to support and construct arguments.

N6.1 Generate terms of a sequence using term-to-term and position-to-term definitions of the sequence.

N6.2 Use linear expressions to describe the nth term of an arithmetic sequence.

AQA Linear specification reference	Learning objectives	Grade	Resource	Common mistakes and misconceptions	Support and homework	
			AQA GCSE Maths Middle sets Student Book; Middle sets Teacher Guide		Middle sets Teacher Guide	Middle sets Practice Book
N6.1	Find the next term in a sequence; Describe the rule for continuing a sequence	E, D	Section 17.1	Expecting all sequences to have common differences. Looking at the first two numbers and assuming that the rest follow this pattern.		Section 17.1
N6.2	Find any term of a sequence given a formula for the nth term; Find the nth term of a linear sequence	E, C	Section 17.2	Mistaking x^2 for $2x$.	GPW 17.2a, b	Section 17.2
N6.2	Find the nth term of a linear sequence; Use the nth term to find terms in a sequence	C	Section 17.3	Not comprehending that when finding the 50th term it is not necessary to find all the terms up to that term.		Section 17.3
N6.2	Find the next few terms in a sequence of patterns; Find the nth term for a sequence of diagrams	E, C	Section 17.4	Not writing down the sequence and trying to do it all mentally. Not making the connection between the structure of the physical pattern and the form the nth term takes.		Section 17.4
N6.1	Find the first few terms of a quadratic sequence by using the nth term; Find the next few terms of a quadratic sequence by looking at differences	D	Section 17.5	Mistaking x^2 for $2x$.		Section 17.5
N6.1	Find the nth term of a simple quadratic sequence	C	Section 17.6	Thinking that $(3x)^2 = 3x^2$. Not checking whether the derived quadratic rule works.		Section 17.6
N5.9	Show step-by-step deduction when proving results	E, D, C	Section 17.7	Not appreciating that a proof shows something works for all values.	GPW 17.7	Section 17.7
N5.9	Show something is false by using a counter-example	C	Section 17.8	Assuming that 'number' means positive whole number. Not identifying an appropriate counter-example.	GPW 17.8	Section 17.8

Chapter 7 Ratio and proportion

Time: 6 hours [F]; 5 hours [H]

N3.1 Use ratio notation, including reduction to its simplest form and its various links to fraction notation.

N3.2 Divide a quantity in a given ratio.

N3.3 Solve problems involving ratio and proportion, including the unitary method of solution.

AQA Linear specification reference	Learning objectives	Grade	Resource – AQA GCSE Maths Middle sets Student Book; Middle sets Teacher Guide	Common mistakes and misconceptions	Support and homework – Middle sets Teacher Guide	Support and homework – Middle sets Practice Book
N3.1, N3.2	Simplify a ratio to its lowest terms Use a ratio in practical situations	E, D	Section 7.1	Swapping over the numbers in the ratio (e.g. 2 : 5 becomes 5 : 2). Simplifying ratios without ensuring the quantities are in the same units.	GPW 7.1a, b	Section 7.1
N3.1, N3.3	Write a ratio as a fraction Use a ratio to find one quantity when the other is known	D, C	Section 7.2	Turning a ratio into a fraction (e.g. the ratio 4 : 5 becomes $\frac{4}{5}$). Failing to find the value of the unit fraction in more complex problems.	GPW 7.2	Section 7.2
N3.3	Write a ratio in the form 1 : n or n : 1	C	Section 7.3	Ignoring different units in a ratio (e.g. simplifying 2 days : 15 hours to 1 : $7\frac{1}{2}$).	GPW 7.3	Section 7.3
N3.3	Share a quantity in a given ratio	D, C	Section 7.4	Forgetting to work out the total number of parts first. Using a ratio as a fraction.	GPW 7.4	Section 7.4
N3.3	Solve word problems involving ratio	C	Section 7.5	Not multiplying both sides of the ratio by the same number. Giving an answer without considering the context.		Section 7.5
N3.3	Understand direct proportion Solve proportion problems, including using the unitary method	D	Section 7.6	Not always seeing the relationships between numbers (e.g. if the cost of 4 items is given, and the price of 8 is asked for).		Section 7.6
N3.3	Work out which product is the better buy	D	Section 7.7	Not making the units the same for each item. Comparing unlike unit rates (e.g. price per gram for one item but amount for 1p for the other).	GPW 7.7	Section 7.7
N3.3	*Solve word problems involving direct and inverse proportion* *Understand inverse proportion*	*D, C, B*	*Section 7.8*	*Dividing by the wrong quantity in conversion problems.*	*GPW 7.8*	*Section 7.8*

© Pearson Education Limited 2010

Chapter 28 Reflection, translation and rotation

Time: 6 hours [F]; 5 hours [H]

G1.6 Recognise reflection and rotation symmetry of 2D shapes.

G1.7 Describe and transform 2D shapes using single or combined rotations, reflections, translations, or enlargements by a positive scale factor and distinguish properties that are preserved under particular transformations.

G5.1 Understand and use vector notation for translations.

AQA Linear specification reference	Learning objectives	Grade	Resource — AQA GCSE Maths Middle sets Student Book; Middle sets Teacher Guide	Common mistakes and misconceptions	Support and homework — Middle sets Teacher Guide	Middle sets Practice Book
G1.7	Draw reflections on a coordinate grid Describe reflections on a coordinate grid	E, D, C	Section 28.1	Drawing the image a different distance from the mirror line than the object. Incorrectly identifying mirror lines parallel to the x- or y-axis.	GPW 28.1	Section 28.1
G1.7, G5.1	Translate a shape on a grid Use column vectors to describe translations	D, C	Section 28.2	Forgetting what the two values in the column vector mean. Using coordinate notation instead of vector notation. Confusing the terms 'transformation' and 'translation'.		Section 28.2
G1.6, G1.7	Draw the position of a shape after rotation about a centre Describe a rotation fully giving the size and direction of turn and the centre of rotation	D, C	Section 28.3	Working out the angle of rotation incorrectly. Turning in the wrong direction. Ignoring the centre of rotation when it is outside the shape.	GPW 28.3	Section 28.3
G1.7	*Transform shapes using more than one transformation* *Describe combined transformations of shapes on a grid*	C	*Section 28.4*	*Not appreciating that two transformations, one followed by another, may be equivalent to a single transformation.* *Not understanding that for 'general' questions where no shape is given, transformations can be carried out on different shapes to investigate what happens.*		*Section 28.4*

Chapter 20 Number skills revisited

Time: 3 hours [F]; 2 hour [H]

N1.3 Understand and use number operations and the relationships between them, including inverse operations and hierarchy of operations.

N1.4 Approximate to a given power of 10, up to three decimal places and one significant figure.

N1.14 Use calculators effectively and efficiently.

N2.1 Understand equivalent fractions, simplifying a fraction by cancelling all common factors.

N2.5 Understand that 'percentage' means 'number of parts per 100' and use this to compare proportions.

N2.7 Calculate with fractions, decimals and percentages.

N3.1 Use ratio notation, including reduction to its simplest form and its various links to fraction notation.

AQA Linear specification reference	Learning objectives	Grade	Resource	Common mistakes and misconceptions	Support and homework	
			AQA GCSE Maths Middle sets Student Book; Middle sets Teacher Guide		Middle sets Teacher Guide	Middle sets Practice Book
N1.3, N1.4, N1.14, N2.1, N2.5, N2.7, N3.1	Understand equivalent fractions		Chapter 20	Forgetting to multiply/divide both the numerator and denominator when finding equivalent fractions and simplifying fractions.		Chapter 20
	Simplify a fraction by cancelling all common factors					
	Recognise that each terminating decimal is a fraction			Applying an incorrect understanding of 'reciprocal'.		
	Convert simple fractions to percentages and vice versa			Writing the ratio in the incorrect order.		
	Use percentages to compare proportions			Forgetting to use BIDMAS when using calculators to perform calculations.		
	Understand 'reciprocal' as multiplicative inverse			Not giving an answer in the context of the problem.		
	Use ratio notation			Treating the digits each side of the decimal point as separate whole numbers , so that 0.95 rounded to 1 d.p. = 0.1.		
	Use brackets and the hierarchy of operations					
	Add, subtract, multiply and divide integers			Dropping zeros when rounding to a number of significant figures (e.g. 5840 = 6 to 1 s.f.).		
	Use calculators effectively and efficiently; use function keys for squares					
	Use inverse operations					
	Round to the nearest integer, to one significant figure and to one, two or three decimal places					
	Give solutions in the context of the problem to an appropriate degree of accuracy					

71

© Pearson Education Limited 2010

N2.5 Understand that 'percentage' means 'number of parts per 100' and use this to compare proportions.

N2.7 Calculate with fractions, decimals and percentages.

N2.7h Including reverse percentage calculations.

AQA Linear specification reference	Learning objectives	Grade	Resource AQA GCSE Maths Middle sets Student Book; Middle sets Teacher Guide	Common mistakes and misconceptions	Support and homework	
					Middle sets Teacher Guide	Middle sets Practice Book
	Number skills: fractions, decimals and percentages (N2.7); using percentages in calculations (N2.7)					
N2.7	Calculate a percentage increase or decrease	D	Section 16.1	Giving the actual increase/decrease as the answer when the amount after the increase/decrease is what is required. Using the multiplier as 1.5 rather than 1.05 for an increase of 5%. Writing '=' between quantities that are not equal, because the '=' sign is used as a shorthand for 'then I do this'.	GPW 16.1	Section 16.1
N2.5, N2.7	Perform calculations involving VAT Perform calculations involving credit Perform simple interest calculations	E, D	Section 16.2	Not seeing that 17.5% = 10% + 5% + 2.5%. Forgetting to add on the initial deposit in credit calculations.	GPW 16.2a, b	Section 16.2
N2.7	Calculate a percentage profit or loss	C	Section 16.3	Confusing cost price and selling price.	GPW 16.3	Section 16.3
N2.7	Perform calculations involving repeated percentage changes	C	Section 16.4	Not understanding when the multiplier should be greater than or less than 1. Using the multiplier as 1.5 rather than 1.05 for an increase of 5%.	GPW 16.4	Section 16.4
N2.7h	*Perform calculations involving finding the original quantity*	*B*	*Section 16.5*	*Not recognising that the problem is not a straightforward percentage increase/decrease question. Not using the correct multiplier.*	*GPW 16.5*	*Section 16.5*

Chapter 25 Quadrilaterals and other polygons

N5.4 Set up and solve simple linear equations.

N6.3 Use the conventions for coordinates in the plane and plot points in all four quadrants, including geometric information.

G1.2 Understand and use the angle properties of parallel and intersecting lines, triangles and quadrilaterals.

G1.3 Calculate and use the sums of the interior and exterior angles of polygons.

G1.4 Recall the properties and definitions of special types of quadrilateral, including square, rectangle, parallelogram, trapezium, kite and rhombus.

AQA Linear specification reference	Learning objectives	Grade	Resource AQA GCSE Maths Middle sets Student Book; Middle sets Teacher Guide	Common mistakes and misconceptions	Support and homework Middle sets Teacher Guide	Middle sets Practice Book
G1.2, N5.4	Calculate the interior angles of quadrilaterals Solve angle problems in quadrilaterals involving algebra	E, D	Section 25.1	Working things out mentally without writing down the calculations.	GPW 25.1	Section 25.1
G1.2, G1.4	Make quadrilaterals from two triangles Use parallel lines and other angle properties in quadrilaterals	E, D	Section 25.2	Giving correct answers but not explaining the properties used.		Section 25.2
G1.3, N5.4	Use the exterior angles of polygons to solve problems Solve more complex angle problems involving exterior and interior angles of a polygon	D, C	Section 25.3	Incorrectly splitting the polygon into triangles.	GPW 25.3a, b	Section 25.3
N6.3	Plot all points of a quadrilateral given geometric information Find the mid-point of a line plotted on the coordinate axes	E	Section 25.4	Plotting the numbers on the x- and y-axes the wrong way round. Not recognising, or be able to name, some of the less common quadrilaterals (e.g. the kite and trapezium). Averaging only the x- or y-coordinate and not both when finding the mid-point.		Section 25.4

© Pearson Education Limited 2010

Chapter 5 Probability

Time: 8 hours [F]; 7 hours [H]

S3.1 Design and use two-way tables for grouped and ungrouped data.

S5.1 Understand and use the vocabulary of probability and the probability scale.

S5.2 Understand and use estimates or measures of probability from theoretical models (including equally likely outcomes), or from relative frequency.

S5.3 List all outcomes for single events, and for two successive events, in a systematic way and derive related probabilities.

S5.4 Identify different mutually exclusive outcomes and know that the sum of the probabilities of all these outcomes is 1.

S5.5h Know when to add or multiply two probabilities: if A and B are mutually exclusive, then the probability of A or B occurring is P(A) + P(B), whereas if A and B are independent events, the probability of A and B occurring is P(A) × P(B).

S5.6h Use tree diagrams to represent outcomes of compound events, recognising when events are independent.

S5.7 Compare experimental data and theoretical probabilities.

S5.8 Understand that if an experiment is repeated, this may – and usually will – result in different outcomes.

S5.9 Understand that increasing sample size generally leads to better estimates of probability and population characteristics.

AQA Linear specification reference	Learning objectives	Grade	Resource AQA GCSE Maths Middle sets Student Book; Middle sets Teacher Guide	Common mistakes and misconceptions	Support and homework Middle sets Teacher Guide	Middle sets Practice Book
S5.1, S5.4	Work out the probability of an event not happening when you know the probability that it does happen	E	Section 5.1	Not remembering that probabilities can be written as fractions, decimals and percentages. Incorrectly subtracting decimals from 1.		Section 5.1
S5.4	Understand and use the fact that the sum of the probabilities of all mutually exclusive outcomes is 1	D	Section 5.2	Adding or subtracting the incorrect values due to misreading the question.	GPW 5.2	Section 5.2
S3.1, S5.3	Understand and use two-way tables	E, D	Section 5.3	Giving an answer that is not requested. Reading the data from the table incorrectly.		Section 5.3
S5.2	Predict the likely number of successful events given the probability of any outcome and the number of trials or experiments	D	Section 5.4	Incorrectly finding fractions of an amount. Not cancelling a fraction to its simplest form.	GPW 5.4	Section 5.4
S5.2, S5.7, S5.8, S5.9	Estimate probabilities from experimental data	C	Section 5.5	Trying to plot decimals worked out to three decimal places or more. Comparing theoretical probability with relative frequency without taking into account the number of trials carried out.		Section 5.5

S5.5h	Calculate the probability of two independent events happening at the same time	C	Section 5.6	Not recognising when a question involves independent events and so adding rather than multiplying the fractions.	GPW 5.6		Section 5.6
S5.6h	*Use and understand tree diagrams in simple contexts*	*B*	*Section 5.7*	*Adding probabilities along the branch rather than multiplying.*			*Section 5.7*

75

© Pearson Education Limited 2010

Chapter 8 Complex calculations

N1.10h Interpret, order and calculate numbers written in standard index form.

N2.7h Including reverse percentage calculations.

AQA Linear specification reference	Learning objectives	Grade	Resource **AQA GCSE Maths Middle sets Student Book; Middle sets Teacher Guide**	Common mistakes and misconceptions	Support and homework	
					Middle sets Teacher Guide	**Middle sets Practice Book**
N2.7h	Perform calculations involving repeated percentage changes	C	Section 8.1	Leaving the multiplier as a percentage, instead of converting to a decimal. Inaccurately converting to a decimal. Not understanding compounding (e.g. treating compound interest as simple).	GPW 8.1	Section 8.1
N2.7h	*Perform reverse percentage calculations*	*B*	*Section 8.2*	*Incorrectly converting percentages to decimals (e.g. 5% = 0.5). Failing to add an increase or subtract a decrease to the given quantity.*		*Section 8.2*
N1.10h	*Interpret and use standard form*	*B*	*Section 8.3*	*Inaccurately converting from a factor of 10 to 10^x and vice versa. Forgetting to include the minus signs in the power for numbers less than 1. Incorrectly entering a number in standard form on a calculator.*		*Section 8.3*

Chapter 18 Linear graphs

Time: 7 hours [F]; 7 hours [H]

N5.4h Including simultaneous equations in two unknowns.

N5.7h Solve linear inequalities in two variables, and represent the solution set on a suitable diagram.

N6.3 Use the conventions for coordinates in the plane and plot points in all four quadrants, including using geometric information.

N6.4 Recognise and plot equations that correspond to straight-line graphs in the coordinate plane, including finding their gradients.

N6.5h Understand that the form $y = mx + c$ represents a straight line and that m is the gradient of the line and c is the value of the y-intercept.

N6.6h Understand the gradients of parallel lines.

N6.11 Construct linear functions from real-life problems and plot their corresponding graphs.

N6.12 Discuss, plot and interpret graphs (which may be non-linear) modelling real situations.

AQA Linear specification reference	Learning objectives	Grade	Resource AQA GCSE Maths Middle sets Student Book; Middle sets Teacher Guide	Common mistakes and misconceptions	Support and homework	
					Middle sets Teacher Guide	Middle sets Practice Book
N6.3	Find the mid-point of a line segment	D, C	Section 18.1	Subtracting the coordinates (instead of calculating an average) when finding the mid-point.		Section 18.1
N6.4	Recognise straight-line graphs parallel to the x- or y-axis Work out coordinates of points of intersection when two graphs cross Plot graphs of linear functions	E, D, C	Section 18.2	Incorrectly calibrating the coordinate axes. Not using a third point as a check when drawing a straight line.	GPW 18.2	Section 18.2
N6.5h, N6.6h	Plot straight-line graphs Find the gradient of a straight-line graph Understand the meaning of m and c in the equation $y = mx + c$ Find the equation of a line	D, C, B	Section 18.3	Forgetting the negative on the gradient.	GPW 18.3	Section 18.3
N6.11	Plot and use conversion graphs	E	Section 18.4	Inaccurately reading from one value on a conversion graph to find another value.		Section 18.4
N6.11, N6.12	Draw, read and interpret distance–time graphs Sketch and interpret real-life graphs	E, D, C	Section 18.5	Drawing and labelling axes before working out the axes range appropriate to the problem.		Section 18.5
N5.4h	*Use a graphical method to solve simultaneous equations*	*B*	*Section 18.6*	*Forgetting to ensure that the individual lines are drawn accurately.* *Not appreciating that the graphical solutions to simultaneous equations are only approximate.*	*GPW 18.6*	*Section 18.6*
N5.7h	*Solve inequalities graphically*	*B*	*Section 18.7*	*Mixing up whether the lines should be dotted or solid.* *Shading the incorrect area.*		*Section 18.7*

© Pearson Education Limited 2010

Chapter 27 3-D objects

Time: 2 hours [F]; 1 hour [H]

G2.4 Use 2D representations of 3D shapes.

AQA Linear specification reference	Learning objectives	Grade	Resource	Common mistakes and misconceptions	Support and homework	
					Middle sets Teacher Guide	Middle sets Practice Book
			AQA GCSE Maths Middle sets Student Book; Middle sets Teacher Guide			
G2.4	Make a drawing of a 3-D object on isometric paper Draw plans and elevations of 3-D objects Identify planes of symmetry of 3-D objects	E, D	Section 27.1	Missing out hidden cubes when converting from a 3-D view to a plan or elevation. Using isometric paper in landscape not in portrait.		Section 27.1

G3.4 Convert measurements from one unit to another.

G3.7 Understand and use compound measures.

AQA Linear specification reference	Learning objectives	Grade	Resource — AQA GCSE Maths Student Book; Middle sets Teacher Guide	Common mistakes and misconceptions	Support and homework — Middle sets Teacher Guide	Middle sets Practice Book
G3.4, G3.7	Convert between different units of area Convert between different units of volume	D, C	Section 30.1	Multiplying by 100 when converting from m^3 to cm^3.	GPW 30.1	Section 30.1
G3.7	Calculate average speeds	D	Section 30.2	Not remembering the formulae. Confusing the decimal parts of an hour with hours and minutes (e.g. using 1 hour 45 minutes as 1.45 hours).	GPW 30.2/30.3	Section 30.2
G3.7	*Make calculations using density*	*C*	*Section 30.3*	*Not remembering the formulae.* *Using the wrong formula.*	*GPW 30.2/30.3*	*Section 30.3*
G3.7	*Recognise formulae for length, area or volume by considering dimensions*	*B*	*Section 30.4*	*Not appreciating that for an expression to represent a quantity every term must have the same dimension.*		*Section 30.4*

© Pearson Education Limited 2010

Chapter 10 Factors, powers and standard form

Time: 3 hours [F]; 6 hours [H]

N1.6 The concepts and vocabulary of factor (divisor), multiple, common factor, highest common factor, least common multiple, prime number and prime factor decomposition.

N1.7 The terms square, positive and negative square root, cube and cube root.

N1.8 Index notation for squares, cubes and powers of 10.

N1.9 Index laws for multiplication and division of integer powers.

N1.9h Fractional and negative powers.

N1.10h Interpret, order and calculate numbers written in standard index form.

AQA Linear specification reference	Learning objectives	Grade	Resource AQA GCSE Maths Middle sets Student Book; Middle sets Teacher Guide	Common mistakes and misconceptions	Support and homework Middle sets Teacher Guide	Middle sets Practice Book
N1.6	Solve problems involving multiples Find lowest common multiples	E, C	Section 10.1	Confusing factors and multiples. Multiplying numbers with a common factor when attempting to find the LCM.	GPW 10.1/10.2	Section 10.1
N1.6	Solve problems involving factors Recognise two-digit prime numbers Find highest common factors	E, C	Section 10.2	Confusing factors and multiples. Missing out 1 as a factor. Forgetting whether it is the highest or lowest common factor that needs to be found. Thinking that 1 is a prime number.	GPW 10.1/10.2	Section 10.2
N1.7, N1.8	Calculate squares and cubes Calculate square roots and cube roots Understand the difference between positive and negative square roots Evaluate expressions involving squares, cubes and roots	E, D, C	Section 10.3	Multiplying by 2 instead of squaring. Writing $\sqrt{36} = -6$ or $\sqrt{36} = \pm 6$ when finding the negative square root.		Section 10.3
N1.8, N1.9h	*Understand and use index notation in calculations* *Understand and use negative powers and numbers to the power of 1 or 0*	*E, B*	*Section 10.4*	*Not being able to describe numbers written with powers.* *Working out 4^5 as 4×5.*	*GPW 10.4/10.6*	*Section 10.4*
N1.6	*Write a number as a product of prime factors using index notation* *Use prime factors to find HCFs and LCMs*	*C*	*Section 10.5*	*Not identifying the prime factors that appear in the decompositions of both numbers when finding the HCF.*	*GPW 10.5*	*Section 10.5*

N1.9, N1.9h, N1.10h	Use laws of indices to multiply and divide numbers written in index notation Carry out calculations with numbers given in standard form	C, B	Section 10.6	Not converting the answers to calculations back into standard form (e.g. $32 \times 10^{-4} \Rightarrow 3.2 \times 10^{-3}$).	GPW 10.4/10.6	Section 10.6

© Pearson Education Limited 2010

Chapter 6 Cumulative frequency

Time: 0 hours [F]; 6 hours [H]

S3.2h Histograms with equal or unequal class intervals, box plots, cumulative frequency diagrams, relative frequency diagrams.

S3.3h Quartiles and inter-quartile range.

S4.4 Compare distributions and make inferences.

AQA Linear specification reference	Learning objectives	Grade	Resource	Common mistakes and misconceptions	Support and homework	
			AQA GCSE Maths Middle sets Student Book; Middle sets Teacher Guide		Middle sets Teacher Guide	Middle sets Practice Book
S3.2h, S3.3h	*Compile a cumulative frequency table and draw cumulative frequency diagrams* *Use cumulative frequency diagrams to analyse data*	*B*	*Section 6.1*	*Mis-reading the graph axes scales.* *Inaccurately summing the frequencies.*		*Section 6.1*
S3.2h	*Draw a box plot from a cumulative frequency diagram*	*B*	*Section 6.2*	*Mis-reading the medians and quartiles.* *Drawing the lower end of the box plot at zero, rather than at the bottom of the lowest class.*		*Section 6.2*
S3.2h, S4.4	*Use cumulative frequency diagrams and box plots to compare data and draw conclusions*	*B*	*Section 6.3*	*Inaccurately plotting cumulative frequency diagrams and box plots.* *Not appreciating the need for a coherent written analysis of diagrams.*		*Section 6.3*

Chapter 29 Circles and cylinders

Time: 7 hours [F]; 6 hours [H]

G1.5 Distinguish between centre, radius, chord, diameter, circumference, tangent, arc, sector and segment.

G2.4 Use 2D representations of 3D shapes.

G4.1 Calculate perimeters and areas of shapes made from triangles and rectangles.

G4.3 Calculate circumferences and areas of circles.

G4.4 Calculate volumes of right prisms and of shapes made from cubes and cuboids.

AQA Linear specification reference	Learning objectives	Grade	Resource AQA GCSE Maths Middle sets Student Book; Middle sets Teacher Guide	Common mistakes and misconceptions	Support and homework	
					Middle sets Teacher Guide	Middle sets Practice Book
G1.5, G4.1, G4.3	Calculate the circumference of a circle Calculate the perimeter of compound shapes involving circles or parts of circles	D, C	Section 29.1	Not multiplying by 2 when the radius is given and the diameter is needed.	GPW 29.1	Section 29.1
G1.5, G4.1, G4.3	Calculate the area of a circle Calculate the area of compound shapes involving circles or parts of circles	D, C	Section 29.2	Multiplying by π before squaring.		Section 29.2
G2.4, G4.1, G4.3, G4.4	Calculate the volume of a cylinder Solve problems involving the surface area of cylinders	C	Section 29.3	Multiplying by π before squaring.	GPW 29.3	Section 29.3

© Pearson Education Limited 2010

Chapter 31 Enlargement and similarity

Time: 2 hours [F]; 5 hours [H]

G1.7 Describe and transform 2D shapes using single or combined rotations, reflections, translations, or enlargements by a positive scale factor and distinguish properties that are preserved under particular transformations.

G1.7h Use positive fractional and negative scale factors.

G1.8 Understand congruence and similarity.

G3.2 Understand the effect of enlargement for perimeter, area and volume of shapes and solids.

AQA Linear specification reference	Learning objectives	Grade	Resource — AQA GCSE Maths Middle sets Student Book; Middle sets Teacher Guide	Common mistakes and misconceptions	Support and homework — Middle sets Teacher Guide	Middle sets Practice Book
G1.7, G3.2	Enlarge a shape on a grid Enlarge a shape using a centre of enlargement	E, D	Section 31.1	Inaccurately counting squares. Adding the scale factor instead of multiplying by the scale factor. Not using the centre of enlargement.	GPW 31.1	Section 31.1
G1.7h, G3.2	Enlarge a shape using a fractional scale factor	C	Section 31.2	Not using the centre of enlargement.	GPW 31.2	Section 31.2
G1.7h, G1.8	Understand similarity and the link with enlargement	B	Section 31.3	Incorrectly simplifying the ratio. Not using corresponding sides when making a ratio.	GPW 31.3	Section 31.3

Chapter 19 Quadratic equations

N5.2h Factorise quadratic expressions, including the difference of two squares.

N5.5h Solve quadratic equations.

AQA Linear specification reference	Learning objectives	Grade	Resource AQA GCSE Maths Middle sets Student Book; Middle sets Teacher Guide	Common mistakes and misconceptions	Support and homework Middle sets Teacher Guide	Middle sets Practice Book
N5.2h	Factorise a quadratic expression that is the difference of two squares	B	Section 19.1	Incorrectly expanding brackets (e.g. expanding $(x - 3)^2$ as $x^2 - 9$).		Section 19.1
N5.2h	Factorise a quadratic of the form $x^2 + bx + c$	B	Section 19.2	Looking for numbers whose sum is c and product b.	GPW 19.2	Section 19.2
N5.2h, N5.5h	Solve quadratic equations by rearranging Solve quadratic equations by factorising	B	Section 19.3	Forgetting that there are two solutions to a quadratic equation.	GPW 19.3	Section 19.3
N5.2h, N5.5h	Write quadratic equations for problems and then solve them	B	Section 19.4	Not identifying which is the unknown value, or using two variables.		Section 19.4

© Pearson Education Limited 2010

Chapter 32 Non-linear graphs

Time: 5 hours [F]; 6 hours [H]

N5.2h Factorise quadratic expressions, including the difference of two squares.

N5.5h Solve quadratic equations.

N6.7h Find the intersection points of the graphs of a linear and quadratic function, knowing that these are the approximate solutions of the corresponding simultaneous equations representing the linear and quadratic functions.

N6.8h Draw, sketch, recognise graphs of simple cubic functions, the reciprocal function $y = 1/x$ with $x \neq 0$, the function $y = k^x$ for integer values of x and simple positive values of k, the circular functions $y = \sin x$ and $y = \cos x$.

N6.11h Construct quadratic and other functions from real life problems and plot their corresponding graphs.

N6.13 Generate points and plot graphs of simple quadratic functions, and use these to find approximate solutions.

AQA Linear specification reference	Learning objectives	Grade	Resource AQA GCSE Maths Middle sets Student Book; Middle sets Teacher Guide	Common mistakes and misconceptions	Support and homework Middle sets Teacher Guide	Middle sets Practice Book
Algebra skills: quadratic expressions (N5.2h, N5.5h)						
N6.11h, N6.13	Draw quadratic graphs Identify the line of symmetry of a quadratic graph Draw and interpret quadratic graphs in real-life contexts	D, C	Section 32.1	Drawing the bottom of the graph flat when a graph has its vertex between two plotted points.	GPW 32.1	Section 32.1
N6.7h	Use a graph to solve quadratic equations	C	Section 32.2	Forgetting to write down all the solutions.		Section 32.2
N6.8h	*Draw cubic graphs Use a graph to solve cubic equations*	*B*	*Section 32.3*	*Incorrectly finding the cube of a negative number. Forgetting to write down all the solutions.*		*Section 32.3*

G2.1 Use Pythagoras' theorem.

AQA Linear specification reference	Learning objectives	Grade	Resource **AQA GCSE Maths Middle sets Student Book; Middle sets Teacher Guide**	Common mistakes and misconceptions	Support and homework	
					Middle sets Teacher Guide	**Middle sets Practice Book**
G2.1	Understand Pythagoras' theorem	C	Section 34.1	Forgetting that x^2 means $x \times x$, not $x \times 2$.		Section 34.1
G2.1	Calculate the hypotenuse of a right-angled triangle Solve problems using Pythagoras' theorem	C	Section 34.2	Forgetting to take the square root to find the final answer. Not correctly identifying the hypotenuse. Forgetting that Pythagoras' theorem only applies to right-angled triangles.		Section 34.2
G2.1	Calculate the length of a shorter side in a right-angled triangle Solve problems using Pythagoras' theorem	C	Section 34.3	Not correctly identifying the hypotenuse. Forgetting to take the square root to find the final answer. Forgetting that Pythagoras' theorem only applies to right-angled triangles. Not identifying the appropriate information when problems are set in context. Not being able to identify the position of the right angle.		Section 34.3
G2.1	Calculate the length of a line segment *AB*	C	Section 34.4	Subtracting instead of adding the two pairs of coordinates.		Section 34.4

87

© Pearson Education Limited 2010

Chapter 33 Constructions and loci

Time: 5 hours [F]; 4 hours [H]

G3.8 Measure and draw lines and angles.

G3.10 Use straight edge and a pair of compasses to do constructions.

G3.11 Construct loci.

AQA Linear specification reference	Learning objectives	Grade	Resource AQA GCSE Maths Middle sets Student Book; Middle sets Teacher Guide	Common mistakes and misconceptions	Support and homework	
					Middle sets Teacher Guide	Middle sets Practice Book
G3.8, G3.10	**Construct perpendiculars** Construct the perpendicular bisector of a line segment **Construct angles of** 90° and **60°** Construct the bisector of an angle	C	Section 33.1	Failing to keep the settings of compasses constant. Rubbing out construction lines. Not using compasses.	GPW 33.1	Section 33.1
G3.11	Construct loci Solve locus problems, including the use of bearings	C	Section 33.2	Confusing a distance from a point with the distance from a line. Making inaccurate constructions. Shading the wrong region.	GPW 33.2	Section 33.2

Chapter 35 Trigonometry

Time: 0 hours [F]; 7 hours [H]

N1.14h Including trigonometrical functions.

G2.2h Use the trigonometrical ratios and the sine and cosine rules to solve 2D and 3D problems.

G3.6 Understand and use bearings.

AQA Linear specification reference	Learning objectives	Grade	Resource AQA GCSE Maths Middle sets Student Book; Middle sets Teacher Guide	Common mistakes and misconceptions	Support and homework Middle sets Teacher Guide	Middle sets Practice Book
G2.2h, N1.14h	Understand and recall trigonometric ratios in right-angled triangles. Know how to enter the trigonometric functions on a calculator	B	Section 35.1	Forgetting that the sine, cosine and tangent ratios only apply to right-angled triangles. Incorrectly using the trigonometric function keys on a calculator.		Section 35.1
G2.2h	Use trigonometric ratios to find lengths in right-angled triangles	B	Section 35.2	Forgetting that the sine, cosine and tangent ratios only apply to right-angled triangles. Not correctly identifying the opposite, adjacent and hypotenuse. Incorrectly using the trigonometric function keys on a calculator.	GPW 35.2	Section 35.2
G2.2h	Use trigonometric ratios to find the angles in right-angled triangles	B	Section 35.3	NOTE: Although sin⁻¹ has been introduced, it is not required at GCSE; therefore, it can be simply said that the inverse is being found. Incorrectly using the trigonometric function keys on a calculator.	GPW 35.3	Section 35.3
G2.2h, G3.6	Use trigonometric ratios and Pythagoras' theorem to solve problems, including the use of bearings	B	Section 35.4	Not identifying the appropriate information when problems are set in context. Drawing a diagram that incorrectly represents the problem. Rounding off values during the intermediate steps of a calculation.		Section 35.4
G2.2h	Solve problems using an angle of elevation or an angle of depression	B	Section 35.5	Not identifying the appropriate information when problems are set in context. Drawing a diagram that incorrectly represents the problem. Rounding off values during the intermediate steps of a calculation.		Section 35.5

© Pearson Education Limited 2010

Chapter 36 Circle theorems

G1.5h Know and use circle theorems.

	Learning objectives	Grade	Resource	Common mistakes and misconceptions	Support and homework	
AQA Linear specification reference			AQA GCSE Maths Middle sets Student Book; Middle sets Teacher Guide		Middle sets Teacher Guide	Middle sets Practice Book
G1.5h	Use chord and tangent properties to solve problems	B	Section 36.1	Giving answers but not explaining the properties used. Not appreciating that listing the unknown facts can help progress the solution to the problem.		Section 36.1
G1.5h	Use circle theorems to solve geometrical problems	B	Section 36.2	Mistaking chords for diameters and therefore incorrectly identifying the subtended angle as 90°.	GPW 36.2	Section 36.2

Lesson Plans

Note: In the Word files on the CD-ROM at the back of this Teacher Guide, fractions have been set using a field code.

To edit a fraction:

1. Click on the fraction to select it.

2. Press **Shift** + **F9** on the keyboard to display the field code.

3. Change the numbers in the fraction by editing in the usual way.

4. With the blinking cursor still in the field code, press **Shift** + **F9** again to display the changed fraction.

1.1 | The data handling cycle

Modular and linear specification reference

S1

Keywords

data, hypothesis

Resources

Guided Practice Worksheet

ActiveTeach Resources

BBC Active video clip

Topic Tutor

Links

Follow up

Middle Practice Book 1.1

Objectives

- Learn about the data handling cycle **D**
- Know how to write a hypothesis **D**

Prior knowledge

Students should know that data is information and that there are many types of data and a range of sources.

Starter

What data would you need to collect if you wanted to find out about how students and teachers travel to school? (Numbers walking, or travelling by bike, bus, car, distances travelled etc.) *Why might someone want to find this out?* (Provision of car parking, bus services, etc.)

Main teaching

- Discuss with the class some examples where we need to collect data, for example:
  - o town planners need to know populations in order to work out how many houses/schools/roads to build
  - o health studies – how do we know if a particular drug is helping to cure a disease?
- Show the explanatory text on the data handling cycle, and remind students of what is meant by a hypothesis. Emphasise that a hypothesis must:
  - o be able to be tested
  - o be about factors that can be measured (not people's opinions)
  - o use clear, unambiguous language.
- Display Examples 1 and 2 using ActiveTeach and discuss.

Common mistakes and misconceptions

Students may have problems formulating a testable hypothesis. They are also likely to think that a hypothesis is not valuable if it is eventually proved false.

Plenary

Split the class into two halves. Ask a student from one half to pose a question and ask students from the other half to formulate at least one hypothesis about it, and to suggest how they could find the data. You may need to start them off, for example: *Is there a link between hours of television watched and students' exam results? Are tall students faster runners?*

1.2 Gathering information

Objectives
- Know where to look for information **D**

Prior knowledge

Students should know that collection and organisation of data is a crucial part of a data handling project.

Starter

Ask the class how they would find data about the following:

- population of a town
- hours of TV watched
- heights of students in Year 11
- probability of throwing two sixes in a row
- theme park prices, etc.

Make a list on the board.

Main teaching

- Display the explanatory text on data sources and Example 3 using ActiveTeach.
  - Explain what is meant by primary and secondary sources.
  - Ask students to identify the sources listed in the Starter as primary or secondary.
- Work through Example 3, and possibly the first one or two questions (if you feel your class needs extra support).

Common mistakes and misconceptions

Students may not be aware that data collected by a third party (even if results of a survey or experiment) is classed as secondary data.

Plenary

Split the class into two halves. Ask a student from one half to suggest a problem for which data needs to be collected (e.g. a timetable, whether a dice is loaded, favourite snacks from the canteen). A student from the other half says how they could find the data, and whether this is primary or secondary.

Modular and linear specification reference
S2.3, S2.4

Keywords
primary, secondary

Resources
none

Guided Practice Worksheet
none

ActiveTeach Resources
BBC Active video clip

Links
none

Follow up
Middle Practice Book 1.2

1.3

Types of data

Objectives
- Be able to identify different types of data **D**

Prior knowledge

Students should know that there are different types of data and that planning is essential for a data handling project.

Starter

Ask students to think of any TV adverts that claim to use numerical data (e.g. 8 out of 10 cats prefer ...). Shampoo and anti-wrinkle creams are usually a rich source of these. Can you measure 'shininess of hair' or 'tastiness of dog food', etc.? Explain that these are examples of qualitative data, but that the manufacturers carry out surveys so that they can say they have proof of their claims.

Main teaching

- Display Example 4 using ActiveTeach. Explain that qualitative refers to a quality (such as taste, colour, etc.) and that quantitative relates to a quantity (a number or measure).

- There is only one type of qualitative data, but quantitative can be further divided into:

  o discrete – data that can be counted, and can only have certain values (numbers of goals scored)

  o continuous – data that can be measured and can fall within a range (heights of people).

- Display and discuss Example 5.

Common mistakes and misconceptions

A possible cause of confusion, rather than a mistake, is that some data can be treated as either discrete or continuous, depending on the context. A good example is age – this is really continuous, but is often treated as discrete, such as when buying child or adult tickets.

Plenary

How would you collect data on exam results, favourite sports, motorbike engine sizes, holiday destinations ...?

Does the type of data help you decide on the way you would collect it?

Modular and linear specification reference

S2.1

Keywords

qualitative, quantitative, discrete, continuous

Resources

Guided Practice Worksheet

ActiveTeach Resources

BBC Active video clip

Links

Follow up

Middle Practice Book 1.3

1.4 Data collection

Objectives
- Work out methods for gathering data efficiently **E**

Prior knowledge

Students should know how to use tally marks.

Starter

Write this sequence of letters on the board:

tjnqmf   dpeft   vtf   mfuuf   gsfrvfodjft

Students will recognise this as a code. Ask how we can work out what the code says, and explain that early secret codes used letter substitution, often just moving each letter on one space in the alphabet. *Why don't we use codes like this now?* (Because they are easy to crack.) We crack them by looking at how often each letter occurs. *What is the most common letter in English?* (Letter **e**.) *What is the most common letter in the code above?* Students might spot that **f** occurs 8 times in 31 letters. If **e** has been replaced by **f**, then the first letter (**t**) might stand for letter **s** and so on. *Who can be first to read the message?* (It says 'Simple codes use letter frequencies'.)

Main teaching

- Ask why we use tally marks, instead of writing a number in the usual way. Students should know that they are useful for collecting ongoing data from experiments or surveys.

- Display Example 6 using ActiveTeach and remind students how to draw up and use a frequency table.
  - o Draw up three columns and fill in the first two columns.
  - o Convert the tally to a frequency.
  - o It is often useful to sum the frequencies to get a total.

Common mistakes and misconceptions

Remind students to take care both when recording and adding tally marks.

Plenary

Ask students to suggest a survey or experiment that could be done at home and which would need a frequency table. Choose some suitable ideas for an alternative homework.

Modular and linear specification reference

S2.4

Keywords

frequency table, tally mark, frequency, frequency distribution

Resources

Guided Practice Worksheet

ActiveTeach Resources

Topic Tutors (×2)

Links

Follow up

Middle Practice Book 1.4

Grouped data

Objectives
- Work out methods for gathering data that can take a wide range of values **D**

Prior knowledge

Students should know the difference between discrete and continuous data.

Starter

Remind students of the inequality symbols, $>$, $<$, $\geq$, $\leq$.

Write some examples on the board. For example:

- *Fill in the blanks with the correct symbol*:

    3 ☐ 5   4.7 ☐ 3.9

- *What values can n take?*

    $3 < n < 7$ $3 \leq n \leq 7$

Main teaching

- Ask students for examples where there might be very large amounts of data (e.g. exam results for a whole year group, sports statistics, national surveys).

    o Explain that to make sense of these examples, the data will need to be put into groups or classes.

- Display Example 7 using ActiveTeach.

    o Show how the tally marks are found.

    o Emphasise that this is discrete data, so groups don't overlap.

    o Adding the frequencies is a good check that you have counted accurately.

    o *What if the data were continuous? For example, if the left-hand column were lengths, where would you put a length of 15.3?*

- Display Example 8.

    o Explain that using the inequality symbols means that we can group continuous data.

Common mistakes and misconceptions

Students sometimes use overlapping class intervals.

Students also convert data to tally marks inaccurately.

Plenary

Ask students to compare the original data lists with the frequency tables. Do the tables give them a clearer picture? (The answer should be yes.)

Modular and linear specification reference
S2.4, S2.5

Keywords
class intervals, grouped data

Resources
none

Guided Practice Worksheet
none

ActiveTeach Resources
Animation

Links
http://weather.lgfl.org.uk/qtables.aspx
This site has large amounts of data about the weather.

Follow up
Middle Practice Book 1.5

1.6 Two-way tables

Objectives
- Work out methods for recording related data **D**

Modular and linear specification reference
S2.5, S3.1

Keywords
two-way table

Resources
none

Guided Practice Worksheet
none

ActiveTeach Resources
Topic Tutor
Using ICT video

Links
none

Follow up
Middle Practice Book 1.6

Prior knowledge

Students should know how to fill in cells in a simple spreadsheet.

Starter

Ask students to volunteer the answers for the Skills check puzzle (A = 12, B = 6, C = 6, D = 10), with the first student to give a correct answer explaining their method to the rest of the class. The row total is 40 and the column total is 24, but there are several acceptable methods.

Main teaching

- Explain that two-way tables are very useful for displaying data. A two-way table makes it easy:
  - o to find missing information
  - o to check accuracy.
- Show how data is recorded in each cell.
- Work through Example 9 using ActiveTeach.

Common mistakes and misconceptions

It is very easy for numerical errors to creep in, so emphasise that it is important to check the totals add up.

Notes on some problem-solving questions

Q3 is starting to develop AO3 skills, as students have to design their own two-way table. If necessary, remind students to look at the structure of the table in Example 9.

Plenary

Ask students to develop their own 3 × 3 puzzle with two or three numbers missing, similar to the Skills check. The first student to create an accurate one could set it for the rest of the class to solve.

1.7 Questionnaires

Objectives
- Learn how to write good questions to find out information **C**

Modular and linear specification reference
S2.3, S2.4

Keywords
survey, questionnaire, response, leading question

Resources
none

Guided Practice Worksheet
none

ActiveTeach Resources
Grade Studio: Problem solving
Topic Tutor

Links
none

Follow up
Middle Practice Book 1.7

Prior knowledge

Students should know a range of methods for recording data.

Starter

Engage students in a short discussion about why questionnaires are useful. *Who might want to use questionnaires?* (Sales people, to find out customer preferences; local councils, to find out what provision to make for roads, schools, car parking etc.; doctors' surgeries, to find out age groups of patients, etc.) Make the point that questionnaires are not usually carried out for fun – they have a purpose.

Main teaching

- Display the Key points and Example 10 using ActiveTeach. Emphasise the need to:
  - decide what you are trying to find out
  - use clear and unambiguous questions
  - plan ahead how to record the information you collect.
- Explain that if time allows, it is a good idea to trial the questionnaire on one or two people before finalising the questions.

Common mistakes and misconceptions

The most likely mistakes are in the design of responses, such as overlapping classes, or gaps between classes. Refer students back to the Key points if you see these mistakes.

Plenary

Ask students to discuss what is wrong with these questions.

1 Do you eat chocolate?

2 How often do you eat chocolate? Tick the correct box.

Never ☐   Occasionally ☐   All the time ☐

3 Tick the correct box.

Chocolate is:

bad for you ☐   very bad for you ☐   OK in tiny amounts ☐

Modular and linear specification reference
S2.2, S2.3, S2.4

Keywords
population, sample, representative, bias, random

Resources
none

Guided Practice Worksheet
none

ActiveTeach Resources
Grade Studio: Knowledge check

Links
none

Follow up
Middle Practice Book 1.8

Objectives
- Know the techniques to use to get a reliable sample **C**

Prior knowledge

Students should know how to write a good questionnaire and how to record data efficiently.

Starter

Ask students to suggest a topic that they would like to find people's opinions about (e.g. *Should break times be extended? Should there be more computer access provided? Should students be allowed to use mobile phones in school?*). (This exercise has more power if you can use a current topic that students might be interested in.) Now ask how long they think it would take to collect and analyse the opinions of everyone in the school. *What can we do to shorten the time it would take?*

Main teaching

- Introduce the terms population, sample, representative, bias and discuss their meanings. Work through Example 11 using ActiveTeach.
  - *How would you make sure a sample is representative?*
  - Introduce the idea of random sampling.
- Ask students what fraction of the school population would be a reasonable sample. Discuss the following sampling methods:
  - just ask Year 10
  - ask all students whose names begin with the letter C
  - ask every tenth person in each class register
  - write all the names down and draw names out of a hat
  - ask only the girls in Years 7 and 8.

Common mistakes and misconceptions

Some students mistake biased samples for random ones.

Plenary

How many people would you ask to take part if you wanted:
- *5% of a population of 3000 (150)*
- *0.1% of a town of 240 000 people (240)*
- *0.05% of 600 000 website users? (300)*
- *0.15% of 2000 village residents (3) Would this be a useful sample?*

Modular and linear specification reference
N2.7

Keywords
fraction key

Resources
calculators

Guided Practice Worksheet
2.1 Fraction of an amount

ActiveTeach Resources
Topic Tutor

Links
none

Follow up
Middle Practice Book 2.1

Objectives

- Find a fraction of an amount with a calculator **E**
- Find a fraction of an amount with a calculator in more complex situations **D**

Prior knowledge

Students should know common metric unit conversions.

Starter

Make sure that students can input fractions onto their scientific calculators. Check that they can read and understand their calculator fraction display.

- Ask students to input $\frac{3}{4}$, and use their calculators to change $\frac{3}{4}$ to 0.75 and back again.
- Ask students to input $1\frac{1}{2}$, and use their calculators to change $1\frac{1}{2}$ to $\frac{3}{2}$ and back again.

Repeat to ensure that all students are using their calculators correctly.

Main teaching

- For basic fractions most scientific calculators are quite user friendly once students know where the fractions button is. A thorough Starter to check students can use their calculators should make this section run smoothly.
- Display Example 1 using ActiveTeach.
  - Allow students to work through parts **a** and **b**.

Common mistakes and misconceptions

Texting thumbs! Using too many fingers or thumbs at once, as many students do when sending a text message on their mobile phones, can result in incorrectly inputting numbers on the calculator. Encourage use of a single finger to make sure no mistakes are made.

Plenary

Ask students to use their calculators to work out some quick-fire questions similar to Q1 and Q2 in the exercise.

Modular and linear specification reference

N2.7

Keywords

quantity

Resources

calculators

Guided Practice Worksheet

2.2 One quantity as a fraction of another

ActiveTeach Resources

Links

Follow up

Middle Practice Book 2.2

Objectives
- Write one quantity as a fraction of another **D**

Prior knowledge

Students should know common metric unit conversions and be able to use common factors to simplify fractions.

Starter

Ask students the following questions:
- *How many students are in the class?*
- *How many of the students are boys?*
- *What fraction of the class are boys?*
- *How many of the students are girls?*
- *What fraction of the class are girls?*

Main teaching

- Display Example 2 using ActiveTeach.
- Emphasise in part **a** that both units must be the same.
- In part **b**, show students, as in the Starter, that the total number must be the denominator.

Common mistakes and misconceptions

A common mistake is not making the denominator the total, but using the 'other number' instead, for example, answering Example 2 part **b** as $\frac{13}{14}$ rather than $\frac{13}{27}$.

Plenary

Ask students to use their calculators to work out some quick-fire questions similar to Q4, Q6 and Q7 in the exercise. Note that the number of days in each month is expected to be known for GCSE.

Objectives
- Use the fraction key on a calculator **E**
- Use the fraction key on a calculator with mixed numbers **D**

Prior knowledge

Students should be able to use a scientific calculator to convert mixed numbers to improper fractions and vice versa, and to simplify fractions.

Starter

Ask students to use their calculators to work out $\frac{9}{10} + \frac{8}{10}$. Some calculators will give $1\frac{7}{10}$ as an answer, and some will give $\frac{17}{10}$. Those with the latter will need to remember to convert their answer to a mixed number. Make sure all students are able to use their calculators to convert mixed numbers to improper fractions and vice versa.

Main teaching

- Most calculator manufacturers have online versions to use. It may be useful to have a couple of different models on screen in the classroom. Students can then work together in groups if access to calculators is limited.
- Display Example 3 using ActiveTeach.
  - Work through the parts and make sure that all students, regardless of the type of calculator they have, get the correct answers.

Common mistakes and misconceptions

Borrowing another person's calculator can be the cause of many incorrect answers!

Notes on some problem-solving questions

The AO2 and AO3 fraction questions in this exercise are testing comprehension of the question rather than higher calculator skills. Students must read the questions carefully and decide which operations (+, −, ×, ÷) need to be carried out on their calculators.

Plenary

Discuss Q4, Q6, Q7 and Q9. Ask students about the way they tackled the questions and how they decided on their particular methods.

Modular and linear specification reference

N1.14

Keywords

Resources

calculators

Guided Practice Worksheet

ActiveTeach Resources

Links

Follow up

Middle Practice Book 2.3

2.4 / Percentages of amounts

Objectives
- Find a percentage of an amount with a calculator **E**
- Find percentages of amounts in more complex situations **D**

Prior knowledge

Students should be able to divide an amount by 2, 4, 10 and 100.
Students should be able to work out 1% and 10% of an amount mentally.

Starter

Ask students to use their calculators to work out the answers to the Skills check questions, explaining different methods if required.

Main teaching

- Display Example 4 using ActiveTeach.
  - Following the Starter, students should not have a problem with this topic, but check that they do understand that percentages such as 7.5% and 150% are possible.

Common mistakes and misconceptions

If some students see a question such as 110% of £300, they leave it blank, as they don't think that percentages over 100% can exist.

Many students treat a percentage such as 0.05% as though it were 5%.

When increasing by a percentage, some students will just add the percentage to the cost, writing, for example, £315 + 15% VAT = £330.

Plenary

Ask students quick-fire questions, without a calculator, such as finding 1%, 10%, 25% and 50% of simple amounts. Extend this to 15% and 75%, and then finish with a couple of harder questions such as 12% of £32.

Modular and linear specification reference

N2.7

Keywords

percent, %

Resources

calculators

Guided Practice Worksheet

ActiveTeach Resources

Grade Studio: Problem solving

Topic Tutor

Links

Follow up

Middle Practice Book 2.4

Objectives

- Write one quantity as a percentage of another **D**
- Write one quantity as a percentage of another in more complex situations **C**

Modular and linear specification reference

N2.7

Keywords

Resources

calculators

Guided Practice Worksheet

ActiveTeach Resources

BBC Active video clips (×2)

Links

Follow up

Middle Practice Book 2.5

Prior knowledge

Students should know common metric unit conversions and be able to convert fractions to decimals with a calculator.

Starter

Ask students to copy and complete:

a $\frac{1}{2} = \frac{\square}{100}$ **b** $\frac{9}{20} = \frac{\square}{100}$ **c** $\frac{12}{25} = \frac{\square}{100}$

Check that students understand that, for example, $\frac{50}{100}$ is 50%, because 'percent' means 'out of 100'.

Main teaching

- Explain that £12 as a percentage of £25 is the same as turning $\frac{£12}{£25}$ or $\frac{12}{25}$ into a percentage. They can either use fractions as in the Starter, $\frac{12}{25} = \frac{48}{100} = 48\%$ or they can use a calculator, $12 \div 25 \times 100 = 48\%$. These skills are needed to solve grade D questions.

- To solve grade C questions, students will often have to use this formula:

  percentage difference $= \frac{\text{actual difference (from start to finish)}}{\text{original value (amount at the start)}} \times 100$

- Display Example 5 using ActiveTeach. Work through the example, noting the following points.

  o In part **a,** 'as a percentage of £25' means 25 is the denominator.

  o In part **b**, notice the units are different.

  o In part **c**, refer to the percentage difference formula above.

Common mistakes and misconceptions

In Q1a, some students will subtract instead of divide, so end up with an answer of £45. Students regularly fail to notice different units in exam questions. Students often do not use the original amount as the denominator when finding a percentage difference.

Plenary

Give students a few non-calculator, easily accessible questions, similar to those in Q1. Then give similar questions that require a calculator. Finish with some questions similar to Q6, including one with a percentage gain.

Modular and linear specification reference

N2.6, N2.7

Objectives

- Calculate a percentage increase or decrease **D**

Prior knowledge

Students should be able to convert percentages to decimals and find a percentage of an amount.

Starter

Ask students to look at the Skills check Q3a. *How much would £200 become if it were:*

- *increased by 5% (£210)*

- *decreased by 5% (£190).*

Do the same for Q3b (£474.60 and £365.40) and Q3c (£389.61 and £276.39).

Main teaching

- Read through the explanatory text on Methods A and B using ActiveTeach.
  - Explain that they used Method A in the Starter, but that they will probably use Method B more often as it tends to be easier.
- Display Example 6.
  - Show students that Method B requires less working.
- Display Example 7.
  - Emphasise how the Method B solutions from Examples 6 and 7 are similar; the only difference is adding to or subtracting from 100%.
- Discuss briefly VAT with students.
- Display Example 8.
  - Ask students which method they now find easier.

Common mistakes and misconceptions

When using Method A, students can forget to add or subtract the percentage they have found.

Plenary

Discuss the two methods and their merits. Ask students which method they prefer and why. Can they think of a type of question where they would actually use the 'other' method?

Keywords

original amount, percentage increase, percentage decrease, reduce

Resources

calculators

Guided Practice Worksheet

ActiveTeach Resources

Animation

Topic Tutor

Links

Follow up

Middle Practice Book 2.6

Modular and linear specification reference

N2.7

Keywords

index number, base, retail prices index

Resources

calculators

Guided Practice Worksheet

2.7 Index numbers

ActiveTeach Resources

Grade Studio: Knowledge check

Links

Follow up

Middle Practice Book 2.7

Objectives

- Understand and use a retail prices index **D**
- Understand and use a retail prices index in more complex situations **C**

Prior knowledge

Students should be able to subtract numbers from 100, and subtract 100 from numbers, and should be able to multiply any number by a fraction with a denominator of 100.

Starter

Ask students these questions:

- *A baguette used to cost £4. It has been reduced in price by 10%. How do you find the new price?* Discuss different methods; show students the quick method of £4 × 90% = £3.60.

- *A different baguette used to cost £3. It has gone up in price by 10%. How do you find the new price?* Discuss different methods; show students the quick method of £3 × 110% = £3.30.

Ask students, using their calculators, to use the quick method to:

- increase and decrease £3.60 by 15% (£4.14, £3.06)
- increase and decrease £6.60 by 5% (£6.93, £6.27).

Main teaching

- Explain what the retail prices index is, and that its main uses are to keep track of inflation, and to help determine pay/income for many people.
- Display Example 9 using ActiveTeach.
    - Work through the examples step by step.
    - Emphasise that the base year amount is always used.

Common mistakes and misconceptions

When students have to find several new prices, they sometimes use the previously found price instead of the base year price.

Notes on some problem-solving questions

In Q5, showing understanding that the price has gone down by $\frac{1}{4}$, or 25%, is sufficient explanation. Students could also show understanding that the new price is $\frac{3}{4}$, or 75%, of the original price.

Plenary

Ask students to discuss the methods they used, and the way they set out their work, in Q1–7.

Modular and linear specification reference

S3.2

Keywords

pie chart, sector, proportion, discrete data

Resources

calculators, protractors

Guided Practice Worksheet

3.1 Drawing pie charts

ActiveTeach Resources

Topic Tutor

Using ICT video

Links

Follow up

Middle Practice Book 3.1

Objectives

• Draw a pie chart **E**

Prior knowledge

Students should be able to express one number as a fraction of another number, and calculate a fraction of a quantity. They should also be able to draw and measure angles and interpret pie charts.

Starter

Ask students to write down all the factor pairs of 360 (1 × 360 = 360, etc.). Ask a set of quick-fire questions of fractions of a quantity.

Main teaching

• Display Example 1 using ActiveTeach.
  o Ask students how they would calculate the angles for the pie chart.
  o Go through either of the two methods. Encourage students to use the method they are more comfortable with.
  o Emphasise the importance of drawing angles accurately. You may have to recap how to use a protractor.
  o Emphasise the importance of labelling the sectors of the pie chart.

Common mistakes and misconceptions

The most common problem is drawing the angles in the pie chart accurately and using the appropriate scale on the protractor.

Often students will measure each angle from the same starting point. It is really worth spending time as a class emphasising that the measurement of each angle is from the end of the previous sector.

Plenary

Discuss which is better when representing data – pictograms, bar charts or pie charts. *Why?*

When is using a pie chart not useful? (For example, when a group has a frequency of zero it does not show up on a pie chart whereas it would on a pictogram or bar chart.)

3.2   Stem-and-leaf diagrams

Objectives
- Draw stem-and-leaf diagrams **D**

Modular and linear specification reference

S3.2

Prior knowledge

Students should be able to order a set of numbers and understand and classify the digits of a set of numbers.

Keywords

stem-and-leaf diagram, key, ascending

Starter

Display these numbers:

19   23   28   2   5   34   41   42   37   37   29

Ask the students to arrange the numbers into groups:

- numbers in the range 0–9 on one line
- numbers in the range 10–19 on the next line below
- numbers in the range 20–29 on the third line below, and so on.

Ask the students to arrange the numbers in each line in ascending order.

Resources

Guided Practice Worksheet

ActiveTeach Resources

Links

Follow up

Middle Practice Book 3.2

Main teaching

- Using the example from the Starter discuss how the information can be represented in a stem-and-leaf diagram.
  - o Draw a stem-and-leaf diagram of the data and emphasise that the stem-and-leaf diagram must be ordered and must have a key.
- Display Example 2 using ActiveTeach.
  - o Discuss Example 2 and show how this will be split into a stem-and-leaf diagram.
  - o Students do Example 2 in pairs.
  - o Emphasise that the stem-and-leaf diagram is much easier to read if the leaf digits are well spaced out.
- Display and work through Example 3.

Common mistakes and misconceptions

Students often forget to include a key and order the leaves.

Encourage students to cross off the numbers when they put them into a stem-and-leaf diagram to ensure they have not missed any out.

Plenary

Place some data on the board, with decimal numbers and draw a stem-and-leaf diagram with two numbers missing, the leaves unordered, and no key. Ask the students to explain why the stem-and-leaf diagram would get no marks.

3.3 Scatter diagrams

Modular and linear specification reference
S3.2, S4.2

Keywords
scatter diagram, correlation

Resources
graph paper

Guided Practice Worksheet
none

ActiveTeach Resources
Topic Tutors (×2)
Using ICT video

Links
none

Follow up
Middle Practice Book 3.3

Objectives
- Draw a scatter diagram on a given grid **D**
- Interpret points on a scatter diagram **D**

Prior knowledge

Students should be able to plot points using coordinates and read scales on axes.

Starter

Draw a coordinate grid and ask students to plot points of your choice. Draw some line segments and mark off different scales, going up in 0.1, or 0.2. Ask for the values of some points on the scale. To extend this, choose some difficult scales to do.

Main teaching

- Display Example 4 using ActiveTeach.
  - o Stress that the scales do not need to start at zero. Explain that when a scale does not start at zero, a squiggle is often (but not always) put on the axis.
  - o Emphasise that care needs to be taken when choosing the scale to make sure points can be plotted accurately.
  - o Show how the points are plotted.
- Ask students to draw the axes and then plot the rest of the points.
  - o *What is the relationship between the maths and science marks?*

Common mistakes and misconceptions

Students often think they have to join all the points with a line instead of just leaving them as plotted points.

When the scales are more difficult, stress that it is worth spending time working out the scale before drawing.

Plenary

Discuss as a class some of the relationships for each of the scatter graphs in Exercise 3D.

Emphasise that when describing a relationship they need to refer to a continuous relationship between the amounts. For example, saying 'as the maths mark increases so does the science mark' is acceptable, whereas 'when the maths mark is good/high the science mark is good/high' is not.

Objectives

- Draw a line of best fit on a scatter diagram **D**
- Describe types of correlation **C**
- Use the line of best fit **C**

Prior knowledge

Students should be able to read scales on axes in a variety of situations and plot a scatter graph accurately.

Starter

Ask students in pairs to describe the relationship between the following:

- the age of a car and its price
- the height and weight of children
- the numbers of hours revising and height
- the numbers of hours revising and the score in a test.

Main teaching

- After taking feedback from students in the Starter, explain that the relationship between variables is known as correlation.
  - Look at each statement in the Starter and ask students to sketch what the scatter graph looks like. Take feedback and explain the three types of correlation.
  - Emphasise that for positive and negative correlation only a line of best fit can be used to predict values.
- Display Example 5 using ActiveTeach. (This builds on Example 4 from the previous lesson.) Draw a line of best fit on the scatter diagram.
- Use the line of best fit to read off and predict some values from the scatter diagram.

Common mistakes and misconceptions

Students often try and make the line of best fit go through the origin, rather than drawing it appropriately.

Notes on some problem-solving questions

In Q3d and Q4d, students must use lines of best fit to make estimated marks. Emphasise that they may need to make predictions without being told to draw a line of best fit, so they must make that decision.

Plenary

Draw three sketches of scatter graphs with positive correlation. Draw a line of best fit in obviously incorrect positions. Students discuss whether they are correct.

Modular and linear specification reference

S4.3

Keywords

line of best fit, linear correlation, positive correlation, negative correlation, no correlation, strong, weak

Resources

graph paper, pencils, rulers, Example 4 from previous lesson

Guided Practice Worksheet

ActiveTeach Resources

Grade Studio: Problem solving

Using ICT video

Links

Follow up

Middle Practice Book 3.4

Objectives
- Draw frequency diagrams for grouped data **D**

Prior knowledge

Students should be able to read data from a table and interpret the class intervals for grouped data.

Starter

Draw a table of continuous data, similar to the height of swimmers in Example 6. Write a list of heights and ask students to decide which class interval they go into. Discuss with reasons what the solutions are.

Main teaching

- Display Example 6 using ActiveTeach.
  - o Emphasise that when the class intervals are the same width the frequency goes on the vertical axis. Remind students that the horizontal axis does not need to start at zero.
  - o Discuss how the frequency diagram is similar to a bar chart but unlike a bar chart there are no gaps since the data is continuous.

Common mistakes and misconceptions

The most common mistake here is the use of grouped labels on the data axes, for example 15–20, rather than the ends of the bar being clearly marked with a 15 at one end and a 20 at the other end.

Plenary

Draw a frequency diagram on the board but start the scale at zero, even though the first group is, for example, 30–40. Have gaps between the bar and an uneven vertical scale. Ask students to work in pairs and say what is wrong with the frequency diagram.

Modular and linear specification reference
S3.2

Keywords
continuous data, frequency diagram

Resources
graph paper

Guided Practice Worksheet
none

ActiveTeach Resources
Animation

Links
none

Follow up
Middle Practice Book 3.5

3.6

Frequency polygons

Objectives

- Draw frequency polygons for grouped data **C**

Prior knowledge

Students should know how to read and plot coordinates.

Starter

Show the students a table similar to the one in Example 7. Write a list of numbers ensuring that some are at the lower and upper bounds of the class intervals, students assign the numbers to the correct class intervals.

Main teaching

- Display Example 7 using ActiveTeach.
  - o For grouped data emphasise finding the mid-point of each group. Show that this is equivalent of finding the mean of the sum of the two end-points. This is the horizontal position to plot.
  - o Emphasise that an extra column for the mid-point is useful to avoid making mistakes.
  - o As a class, draw the frequency polygon. The horizontal axis should always be a continuous scale, regardless of whether grouped or ungrouped data.

Common mistakes and misconceptions

The most common mistake is putting a grouped label on the horizontal axis rather than a continuous scale. Students frequently plot the upper bound instead of the mid-point.

Plenary

Design a poster showing the key information on frequency polygons or discuss the answer to Exercise 3G Q5. Write down the key points of comparison of the frequency polygons.

Modular and linear specification reference

S3.2, S4.4

Keywords
frequency polygon, continuous data, mid-point

Resources
graph and squared paper

Guided Practice Worksheet
3.6 Frequency polygons

ActiveTeach Resources
Animation
Grade Studio: Knowledge check
Using ICT video

Links
none

Follow up
Middle Practice Book 3.6

Modular and linear specification reference

S3.3, S4.1

Keywords

average, mean, median, mode, modal value, range

Resources

sticky notes

Guided Practice Worksheet

4.1 Averages and range

ActiveTeach Resources

Animations (×2)

BBC Active video clips (×2)

Topic Tutor

Links

http://www.bbc.co.uk/skillswise/numbers/handlingdata/numericalanalysis/mean/flash0.shtml

A flash activity exploring the mean.

http://www.quia.com/rr/51667.html

An interactive quiz on mean, median, mode and range.

Follow up

Middle Practice Book 4.1

Objectives

- Find the mean, median and range from a set of data, including data given in a stem-and-leaf diagram **E, D**

Prior knowledge

Students should be able to order whole numbers and decimals. They should also be able to find the number half way between two numbers.

Starter

Display a 0–10 number line on the board. Write decimals between 0 and 10 on sticky notes and ask students to place them in the correct positions on the number line.

Main teaching

- Introduce averages and range.
  - o Explain that there are three different types of average. Explain that the median tells you the middle value and the mode tells you the most common value. Explain that the mean is the only average which takes into account all the values.
  - o Explain that the range is a measure of how spread out the data is.
- Gather primary data from the class.
  - o Choose 10 students and ask them how many brothers or sisters they have. Record the results in a list on the board. Ask students to calculate the mean, median, mode and range of the data. Revise methods of calculating averages and range if necessary.
  - o Highlight the fact that the median is the $\left(\frac{n+1}{2}\right)$th value, and explain that with an even number of data values there are two middle values, and that the median is half way between them.
- Display Example 1 using ActiveTeach and work through the parts step by step.

Common mistakes and misconceptions

Students often omit units when writing averages or range.

Plenary

Calculate the mean, median, mode and range of the word lengths in a sentence (for example, the first 'Why learn this' sentence using ActiveTeach).

Objectives

- Calculate the mode, median and range from an ungrouped frequency table **E**

Prior knowledge

Students should be able to interpret data presented in a frequency table.

Starter

Ask all students to stand. Call out 2-digit numbers and ask students to mentally add each number to the total. Students could display their totals on mini-whiteboards after every five numbers. Students with the incorrect total are eliminated and must sit down. Increase the speed until only one student remains.

Main teaching

- Gather primary data for use in the lesson.
  - Ask students how many brothers and sisters they have and record the class results as a list and in a tally chart.
  - Fill in the frequency column and ask students to calculate the total frequency. *What does the sum of the frequencies represent?*
  - Explain that the data value with the highest frequency is the mode.
  - *What is the range of this data?*
  - Demonstrate the use of the formula: median $= \left(\frac{n+1}{2}\right)$th value.
  - Demonstrate counting through the frequencies to find the position of the median value.
- Display the frequency table from Example 2 using ActiveTeach.
  - *How many times was the dice rolled?*
  - Ask students to use the formula to find the position of the median.

Common mistakes and misconceptions

Students frequently confuse the frequencies and the data values. Reinforce the fact that the frequency tells you the number of times a certain data value occurs.

Plenary

Use the data gathered in the lesson to calculate the probability that a student chosen at random will have no brothers or sisters. *Which average tells us the most likely number of brothers or sisters for a student chosen at random?*

Modular and linear specification reference
S3.3, S4.1

Keywords
ungrouped frequency table

Resources
mini-whiteboards (optional)

Guided Practice Worksheet
none

ActiveTeach Resources
none

Links
none

Follow up
Middle Practice Book 4.2

Objectives

- Calculate the mean from an ungrouped frequency table **D, C**

Prior knowledge

Students should be able to multiply 2-digit numbers and decimals by single-digit numbers without a calculator. They should also be able to multiply and divide confidently with a calculator and round their answers when appropriate.

Starter

Display a multiplication grid on the board. Choose two students to play 'Three in a row'. Students select squares in turn then have, for example, 20 seconds to mentally complete the multiplication for that square. If they are successful they 'win' the square.

×	2.1	73	25	0.15
4				
8				
9				
6				

The first player to win three squares in a row (horizontally, vertically or diagonally) is the winner.

Main teaching

- Gather primary data to use in the lesson.
  - Roll a dice 20 times (or use a dice-rolling simulator with the electronic whiteboard). Record the results as a list and in a frequency table.
  - Find the mean from the list and explain how you can find the mean more quickly from the frequency table by adding an extra column to show the total number of times each number was rolled.
  - *How many times did we roll a 6? What was the total of these rolls?*
  - Explain that the total of the extra column is the sum of all the rolls.
- Demonstrate the method using the data in Example 3 using ActiveTeach.

Common mistakes and misconceptions

Students will frequently divide by the number of rows in the frequency table (i.e. the number of different data values) and not by the sum of the frequencies. *How many data values are there in total?*

Notes on some problem-solving questions

Q3 is an extended question which will take up to 20 minutes to complete. Students should draw three separate frequency tables and add a column to each for 'Number of gold stars × frequency'.

Plenary

Choose frequency tables used in the lesson and ask students to calculate the mode, median and range of the data.

Modular and linear specification reference

S3.3, S4.1

Keywords

ungrouped frequency table

Resources

dice or dice simulator

Guided Practice Worksheet

ActiveTeach Resources

Grade Studio: Problem solving

Links

http://www2.whidbey.net/ohmsmath/webwork/javascript/diceroll.htm

A simple simulator for rolling a dice.

Follow up

Middle Practice Book 4.3

Objectives

- Find the modal class from a grouped frequency table **D**
- Estimate the range from a grouped frequency table **D**
- Work out the class interval which contains the median from data given in a grouped frequency table **C**

Prior knowledge

Students should understand the difference between discrete and continuous data and be able to use a grouped frequency table. They should also know the meaning of the symbols $<$, $>$, $\leq$ and $\geq$.

Starter

Find the median, mode and range of these numbers:

   3   2   4   5   6   2   2   4   7

Main teaching

- Remind students that continuous data is often presented in a grouped frequency table. *Give me an example of continuous data. Give me an example of discrete data.*

- Display the grouped frequency table from Example 4 using ActiveTeach.

   o Explain that you cannot calculate the exact median, mode or range because you do not know any of the data values exactly.

   o Explain that you can find an estimate for the range and work out which class interval contains the median and mode.

   o *Which class interval has the highest frequency?* ($5 \leq h < 10$) Explain that this class interval is the mode, or modal class interval.

   o *What is the largest possible data value?* (< 25) *What is the smallest possible data value?* (5) Explain that an estimate for the range is the largest possible value minus the smallest possible value.

- Revise finding the median from an ungrouped frequency table. Explain that the same method allows you to find the class interval which contains the median in a grouped frequency table.

Common mistakes and misconceptions

Students should remember that statistics calculated from grouped frequency tables are estimates. They should understand that the estimate for the range is an upper limit.

Plenary

Discuss the choice of suitable class intervals for a tally chart for different surveys and experiments (e.g. heights of students, reaction times).

Modular and linear specification reference

S3.3, S4.1

Keywords

class interval, grouped frequency table, modal class, estimate

Resources

Guided Practice Worksheet

4.4/4.5 Finding the range and averages from a grouped frequency table (combined)

ActiveTeach Resources

Topic Tutor

Links

Follow up

Middle Practice Book 4.4

Modular and linear specification reference
S3.3, S4.1

Keywords
estimate, mid-point

Resources
none

Guided Practice Worksheet
4.4/4.5 Finding the range and averages from a grouped frequency table (combined)

ActiveTeach Resources
Grade Studio: Knowledge check
Topic Tutor

Links
http://douis.net/Iiws/groupfreq1.htm
Four interactive screens allowing students to enter values to find estimates of means and check their answers.

Follow up
Middle Practice Book 4.5

Objectives

- Estimate the mean of data given in a grouped frequency table **C**

Prior knowledge

Students should be familiar with grouped frequency tables. They should be able to find the value half way between two whole numbers or decimals. They should also be able to multiply and divide confidently, and round answers when appropriate.

Starter

Secretly choose a number correct to one decimal place such as 12.8. Ask students to guess your number, and tell them if they are too high or too low. Encourage students to use the mid-point of their upper and lower bounds as their next guess.

Main teaching

- Display the grouped frequency table in Example 5 using ActiveTeach.
  - Explain that it is not possible to write the list of data values exactly from the frequency table. *Can you calculate the exact mean?* (No)
  - Explain that to find an estimate of the mean you have to assume that all the values in a group are in the middle of that group.
  - Explain that you can find the mid-point of a group by adding together the maximum and minimum values from that group and dividing by 2.

Common mistakes and misconceptions

Students often calculate the mid-points of class intervals incorrectly for grouped discrete data (i.e. the mid-point of the class interval 10–19 is 14.5, not 15).

Plenary

Discuss the effect the choice of class interval has on the accuracy of the estimate for the mean.

Do you think bigger class intervals will produce a more accurate estimate?

5.1

Probability that an event does not happen

Objectives

- Work out the probability of an event **not** happening when you know the probability that it does happen **E**

Prior knowledge

Students should be able to subtract simple fractions and decimals from 1, and subtract whole numbers from 100. They should also be able to work out fraction, decimal and percentage equivalents.

Starter

Divide the class into two. Allow half the class to use a calculator and the other half have to work mentally. Give the class a mixture of easy and hard quick-fire questions of the form '1 – an amount'. After 10 or 12 questions, ask the class to swap between using a calculator and working mentally. Ask another 10 or 12 questions.

Discuss mental strategies, especially when answering questions with fractions, and which types are easier with a calculator.

Main teaching

- Display Example 1 using ActiveTeach.
  - Emphasise to students that showing workings (such as those in the yellow boxes on the right-hand side) will save marks if they make careless mistakes in the exam.

Common mistakes and misconceptions

Many students have difficulty with questions such as Q3 where they have to work out $1 - 0.05$.

When students attempt questions such as Q5 and Q6, many get confused and think there is a link between the numbers and the colours.

Plenary

Repeat the Starter, but this time put the amounts in context so that the students realise that these types of questions all involve the same type of skill.

Modular and linear specification reference

S5.1, S5.4

Keywords

probability, event, not

Resources

Guided Practice Worksheet

ActiveTeach Resources

Animations: Before you start this chapter (×2)

BBC Active video clips: Before you start this chapter (×2)

Links

Follow up

Middle Practice Book 5.1

5.2 Mutually exclusive events

> **Objectives**
> - Understand and use the fact that the sum of the probabilities of all mutually exclusive outcomes is 1 **D**

Prior knowledge

Students should be able to add simple fractions and decimals, and be able to subtract simple fractions and decimals from 1.

Starter

Put 5 blue, 3 red and 2 green counters into a bag or box. Make sure that the students are watching. Ask a series of questions relating to the probability of picking out one counter at random, such as:

What is the probability that the counter is:

   a *blue*   **b** *not blue*   **c** *not red*   **d** *blue or red*   **e** *not red or green?*

Add 1 white counter to the bag/box and ask students to write out six probability questions and answers relating to the counters in the bag/box.

Main teaching

- Discuss mutually exclusive events. This is a difficult concept for many students.
    - Start by tossing a coin and ask for possible outcomes (heads or tails) and probabilities, then ask for the sum of all the probabilities.
    - Repeat with a 4-sided dice, if available, then a 6-sided dice.
- Return to the bag/box used in the Starter. Ask for the probability of green and the probability of not green. Then ask for the sum of the probabilities of green and not green. Explain that the sum = 1 as they are mutually exclusive events, that is, they cannot happen at the same time.
- Display Example 2 using ActiveTeach.
    - Work through both parts step by step.

Common mistakes and misconceptions

Students often don't read questions carefully enough and so add or subtract incorrect values.

Plenary

In pairs, ask the students to explain to each other what a mutually exclusive event is. Discuss what skills are required to tackle these types of questions.

Modular and linear specification reference

S5.4

Keywords

mutually exclusive, or, add, certain

Resources

blue, red, green and white counters, bag or box (for Starter)

coin, 4-sided dice, 6-sided dice (for Main teaching)

Guided Practice Worksheet

5.2 Mutually exclusive events

ActiveTeach Resources

Using ICT video

Links

Follow up

Middle Practice Book 5.2

Two-way tables

Objectives

- Understand and use two-way tables **E, D**

Prior knowledge

Students should understand basic probability.

Starter

Copy this table (or use a similar one) onto the board.

	Football	Netball	Basketball	Rugby	Athletics	Total
Male						
Female						
Total						

Complete the table by asking students their favourite sport.

Ask students to work out the probability of selecting one person at random from different groups of people, such as girls who play netball, boys who play rugby, etc.

Choose a straightforward multiple of the number of students in the class, and ask how many would be expected out of this total to be in each category of the table.

Main teaching

- Display Example 3 using ActiveTeach.
  - Ignore the questions initially and ask the students to give facts about the offices, such as 7 offices have 2 windows and 1 door, etc.
  - Point out that a two-way table often, but not always, has totals.
  - It doesn't matter whether you add the data horizontally or vertically, the overall total is the same.
  - Work through the questions step by step with special emphasis on the different skills required in parts **bi** and **bii**.

Common mistakes and misconceptions

The usual mistake made by students in exams is not reading a question carefully and so giving information not requested (e.g. in Exercise 5C Q1di, writing $\frac{6}{15}$ rather than $\frac{15}{40}$).

Plenary

100 people flipped a coin. Some of their results are shown in the table.

	Heads	Tails	Total
Girls	21		45
Boys			
Total	58		

Ask students to fill in the missing entries in the table.

Modular and linear specification reference

S3.1, S5.3

Keywords

two-way table

Resources

Guided Practice Worksheet

ActiveTeach Resources

Topic Tutor

Using ICT video

Links

Follow up

Middle Practice Book 5.3

5.4 Expectation

Objectives
- Predict the likely number of successful events given the probability of any outcome and the number of trials or experiments **D**

Prior knowledge

Students should understand basic probability and be able to find a fraction of an amount.

Starter

Ask students questions relating to basic probability using coins, dice and playing cards.

- *What is the probability of flipping heads?*
- *What is the probability of rolling a 3?*
- *What is the probability of turning over a Jack of Hearts?*

Main teaching

- Display Example 4 using ActiveTeach.
  - Work through parts **a** and **b**.
  - Ask similar questions such as:
    With 30 rolls how many 1s would you expect? (5)
    With 60 rolls how many 2s would you expect? (10)
    With 90 rolls how many odd numbers would you expect? (45)
    After how many rolls would you expect to get three 3s? (18)

Common mistakes and misconceptions

Some students have difficulty in finding fractions of an amount, with or without a calculator.

Plenary

Ask students a few fairly straightforward questions such as:

If the probability that [student's name] can throw a peanut in the air and catch it in their mouth is 0.8, how many peanuts are likely to be on the floor if they try 200 times?

Modular and linear specification reference
S5.2

Keywords
likely, estimate, trial

Resources
none

Guided Practice Worksheet
5.4 Expectation

ActiveTeach Resources
Grade Studio: Problem solving
Topic Tutor

Links
none

Follow up
Middle Practice Book 5.4

Objectives
- Estimate probabilities from experimental data **C**

Prior knowledge

Students should be able to convert a fraction to a decimal and be able to use a calculator.

Starter

Ask students the following questions:

- *What is the probability of flipping a coin and getting heads?*
- *If you flipped a coin 10 or 100 or 1 million times, would you always get half heads and half tails? Why not?*
- *How could you work out the probability of a drawing pin landing point up when it is dropped?*

Main teaching

- Display Example 5 using ActiveTeach.
  - Work through each part step by step.
  - Ensure that students understand that plotting points to three decimal places is a 'near enough' exercise, and that part **b** is an 'educated guess' (unless many more trials are carried out).

Common mistakes and misconceptions

Students who try to plot decimals to three decimal places or more often make mistakes; they should round sensibly before trying to plot.

Students sometimes compare theoretical probability with relative frequency without taking into account the number of trials carried out.

Plenary

Ask students the following questions:

- *How many times do you think you should flip a coin to check if it is fair or biased?*
- *How many times do you think you should roll a 6-sided dice to check if it is fair or biased?*
- *How many times do you think you should spin a 10-sided spinner to check if it is fair or biased?*

If there are any differences in the number of trials suggested, ask students to give their reasons.

Modular and linear specification reference

S5.2, S5.7, S5.8, S5.9

Keywords

theoretical probability, experimental probability, estimated probability, relative frequency, successful trials, expect

Resources

Guided Practice Worksheet

ActiveTeach Resources

Animation

BBC Active video clip

Topic Tutor

Links

Follow up

Middle Practice Book 5.5

Objectives

- Calculate the probability of two independent events happening at the same time **C**

Modular and linear specification reference

S5.5h

Keywords

independent, multiply, and

Resources

10p, 50p and £1 coin, red and blue dice

Guided Practice Worksheet

5.6 Independent events

ActiveTeach Resources

Topic Tutor

Links

Follow up

Middle Practice Book 5.6

Prior knowledge

Students should be able to add and multiply fractions.

Starter

Ask students the following questions involving basic probability:

- *If I flip a 10p coin, how many possible outcomes are there?*
- *If I flip a 10p coin and a 50p coin, how many possible outcomes are there?*
- *If I flip a 10p coin, a 50p coin and a £1 coin, how many possible outcomes are there?*
- *If I roll a red dice, how many possible outcomes are there?*
- *If I roll a red dice and a blue dice, how many possible outcomes are there?*
- *How many possible outcomes are there if I flip a coin and roll a dice?*

Main teaching

- It is essential that students are comfortable with multiplying two fractions together.
- Discuss independent events, relating the discussion back to the Starter. This could also be a convenient time to remind students of mutually exclusive events from section 5.2.
- Display Example 6 using ActiveTeach.
  - Work through the example step by step, relating the denominators back to the answers in the Starter.

Common mistakes and misconceptions

Students may not recognise when a question involves independent events and so often add rather than multiply the fractions.

Plenary

Ask students to define independent events and mutually exclusive events.

Ask the following questions:

- *If two events A and B are independent, what can be said about P(A and B)?*
- *If two events A and B are mutually exclusive, what can be said about P(A and B)?*
- *Give two events which are/are not independent.*
- *Give two events which are/are not mutually exclusive.*

5.7 Tree diagrams

5.7

Objectives
- Use and understand tree diagrams in simple contexts **B**

Prior knowledge

Students should be able to add and multiply fractions, and be able to add and multiply decimals.

Starter

Ask students how many possible outcomes there are when flipping two coins. Draw a tree diagram to show all the possibilities and discuss the various probabilities, especially 'one head and one tail'.

Ask students how many possible outcomes there are when flipping three coins. Add the third coin to the tree diagram and discuss the various probabilities, especially 'only one tail', or 'two tails and one head'.

Main teaching

- Display Example 7 using ActiveTeach.
  - o Relate the 'one of each colour' back to the 'one head and one tail' in the Starter.
  - o Emphasise the importance of paying special attention to the headings and the labelling of the branches of the trees.
  - o Focus on the tree diagram. Be clear as to where the probabilities are written, and how the outcomes such as RR are calculated.
- Display Example 8.
  - o Work through the example step by step with constant reference back to the tree diagram above.

Common mistakes and misconceptions

Students often struggle with multiplication and addition of fractions.

Plenary

Ask students to look again at Exercise 5H Q7. Then display Example 6, focusing on the sample space diagram. Discuss the types of questions where one diagram is better than the other. Can students suggest types of questions where a tree diagram is better than a sample space diagram, and vice versa?

Modular and linear specification reference
S5.6h

Keywords
tree diagram, combined events

Resources
none

Guided Practice Worksheet
none

ActiveTeach Resources
Grade Studio: Knowledge check
Topic Tutor

Links
none

Follow up
Middle Practice Book 5.7

Modular and linear specification reference

S3.2h, S3.3h

Objectives

- Compile a cumulative frequency table and draw cumulative frequency diagrams **B**
- Use cumulative frequency diagrams to analyse data **B**

Keywords

lower quartile, upper quartile, inter-quartile range, median, cumulative frequency

Prior knowledge

Students should know how to read graphs accurately.

Resources

Starter

What is a running total? Add these numbers mentally: 1 4 7 12 5 6 2

Add these using a calculator: 23 38 55 60 41 33

In what situations might we want to use running totals? Where might we make mistakes?

Guided Practice Worksheet

ActiveTeach Resources

Animations (×2)

Topic Tutor

Using ICT video

Main teaching

- Go through the definitions and Example 1 using ActiveTeach.
  - Explain that the first step is to add a column to the table for the running total (the cumulative frequency).
  - Show how to plot the points, and having drawn the curve, how to find the median. Students will be familiar with the median and have previously found the median class from a frequency table. *How does the median value compare with the median class?* (3.6 minutes, compared with the 3–4 class.) So, using a cumulative frequency curve, the median can be found with more precision.
  - Explain how to find the quartiles and the inter-quartile range.
- Students could now answer Exercise 6A, or alternatively, go through Example 2 before asking students to tackle both Exercises 6A and 6B.

Links

Follow up

Middle Practice Book 6.1

Common mistakes and misconceptions

Students may mis-read the graph axes scales. Inaccurate summing of frequencies will lead to problems.

Plenary

Cumulative frequency diagrams are usually S-shaped. These curves are called *ogives* (pronounced *o-jives*). *Why are the curves shaped like this?* Encourage students to interpret factors causing this shape – for example, *How many adults have heights less than, say, 4 feet 6 inches? How many are between 4 feet 6 inches and 6 feet? How many are over 6 feet?* The bell curve arising from most natural distributions gives rise to an S-shaped cumulative frequency curve. You could ask students for other examples that they could test (e.g. lengths of index fingers, wrist circumference, foot length, lengths of leaves from an oak tree, etc.).

Objectives
- Draw a box plot from a cumulative frequency diagram **B**

Modular and linear specification reference
S3.2h

Keywords
box plot, whiskers

Resources
none

Guided Practice Worksheet
none

ActiveTeach Resources
Animation
Topic Tutor
Using ICT video

Links
none

Follow up
Middle Practice Book 6.2

Prior knowledge

Students should know how to find the median, quartiles and inter-quartile range.

Starter

Suggest a range of times taken in minutes to travel to school:

23 45 13 56 21 15 49 45 33 23 18 22 30 29

How easy is it to get a picture of this distribution? (Not very easy, and would be much harder with a larger set of results.) *How can we make it easier?* (Find the median, quartiles, etc.) *What must we do first?* (Put the numbers in order.)

Main teaching

- Show the explanatory text on box plots using ActiveTeach.
  - *What does the box plot tell you about the data?* Students should appreciate that the box plot shows how the data is spread, but doesn't tell them any details, not even the sample size.

- Compare the box plot in the explanatory text with those in Examples 3 and 4.
  - The shapes show that, in Example 3, for instance, the data is fairly evenly spread, whereas in Example 4, there are more values towards the lower end.
  - The two extreme values are also shown. Point out, however, that these might be estimates.

- It might be useful for students to draw their box plots in Exercise 6C immediately below their answers for Exercises 6A and 6B, if space allows.

Common mistakes and misconceptions

Mis-reading the medians and quartiles will cause problems

Students may draw the lower end of the box plot at zero, rather than at the bottom of the lowest class.

Plenary

Ask if it is possible to draw box plots directly from the raw data. *What are the advantages of this?* (The ends of the whiskers will have actual rather than estimated values.) *What are the disadvantages?* (The data still has to be processed in order to find the median and quartiles.)

Objectives

- Use cumulative frequency diagrams and box plots to compare data and draw conclusions **B**

Prior knowledge

Students should know how to find the median, quartiles and inter-quartile range.

Starter

What averages can you find for this set of data?

2   2   9   6   2   9   3   5   4   0   2   6   1   2   7

(Mode = 2; median = 3; mean = 4)

Main teaching

- Show the data for Example 5 using ActiveTeach. Explain that there are often situations where we want to compare two or more large sets of data.

  o *How could you compare the two types of battery?* Students might suggest finding an average, in which case, probe to find which kind of average – mode, median or mean? They are all valid – the mode is easy to find here (28–32 for Superamp and 24–28 for Powerplus) but a mean would have to be an estimated mean.

  o *What else can you do with the data sets?* Establish that cumulative frequency diagrams and box plots can be plotted. Show that these immediately give a much clearer visual picture of how the two types compare.

  o Explain that we can go further than an image, by calculating values – the median, upper and lower quartiles and the inter-quartile range.

  o Emphasise that the inter-quartile range also gives a measure of the consistency, or reliability.

Common mistakes and misconceptions

Plotting cumulative frequency diagrams and box plots inaccurately will cause difficulties. Students also have difficulty in writing a coherent analysis, and may need support to do this.

Plenary

What are the advantages of presenting data as box plots? (It is easy to see the medians, quartiles and inter-quartile range, and it is easy to make comparisons.) *What are the disadvantages?* (The raw data, extreme values and the sample size may not be shown.)

Modular and linear specification reference

S3.2h, S4.4

Keywords

Resources

Guided Practice Worksheet

ActiveTeach Resources

Grade Studio: Knowledge check

Grade Studio: Problem solving

Topic Tutor

Links

Follow up

Middle Practice Book 6.3

Objectives
- Simplify a ratio to its lowest terms **E**
- Use a ratio in practical situations **D**

Prior knowledge

Students should know and be able to use common metric unit conversions.

Starter

Skills check Q2 may not be enough practice of metric unit conversions for some students, so extra quick-fire questions may be beneficial.

Ask students to simplify the following fractions to their lowest terms:

a $\frac{3}{6}$ $(=\frac{1}{2})$ **b** $\frac{15}{25}$ $(=\frac{3}{5})$ **c** $\frac{100}{75}$ $(=\frac{4}{3})$ **d** $\frac{3000}{600}$ $(=\frac{5}{1})$

Main teaching

- Demonstrate how to simplify the following fractions to their lowest terms:

$$\frac{1\frac{1}{2}}{3} \left(=\frac{3}{6}=\frac{1}{2}\right) \quad \text{and} \quad \frac{1\frac{1}{2}}{7\frac{1}{2}} \left(=\frac{5}{15}=\frac{1}{3}\right)$$

- Display Example 1 using ActiveTeach.
  - Emphasise that the ratio notation ':' means 'to'.
  - Show how similar the methods are for simplifying ratios and fractions, and explain that students are always expected to give ratios in their simplest forms.
- Display the explanatory text describing ratios and display Example 2.
  - Work through parts **a** and **b** step by step; some students may prefer to use the method shown in part **b** for part **a** as well.

Common mistakes and misconceptions

Students may swap over the numbers in the ratio, so 2 : 5 becomes 5 : 2. The most common mistakes are made when simplifying ratios involving different unit measurements such as 2 m : 50 cm becomes 1 : 25.

Plenary

Ask students to simplify the following orange squash : water ratios:

a 9 : 3 **b** 25 : 5 **c** 5 : $1\frac{1}{4}$

Which ratio has the highest proportion of orange squash?

Modular and linear specification reference
N3.1, N3.2

Keywords
ratio, simplify, maps, scale drawing

Resources
none

Guided Practice Worksheets
7.1a Simplifying ratios
7.1b Using ratios

ActiveTeach Resources
BBC Active video clips (×2)
Topic Tutors (×2)

Links
none

Follow up
Middle Practice Book 7.1

7.2 Ratios and fractions

Objectives
- Write a ratio as a fraction **D**
- Use a ratio to find one quantity when the other is known **D, C**

Prior knowledge

Students should be able to carry out basic addition and subtraction of fractions.

Starter

Ask students questions such as:
- If I eat $\frac{1}{5}$ of a cake, how much is left?
- If $\frac{1}{3}$ of a pizza weighs 200 g, how much does the whole pizza weigh?
- If $\frac{3}{4}$ of a pizza weighs 200 g, how much does the whole pizza weigh?

Main teaching

- A cake is cut into 7 equal pieces – 2 pieces are eaten, and 5 pieces are not eaten. What is the ratio of eaten : not eaten? (2 : 5)
  Write 2 : 5 on the board and remind students that there are 7 pieces of cake altogether.
  - What fraction has been eaten? What fraction hasn't been eaten?
- Display Example 3 using ActiveTeach.
  - Work through part **a**, reminding students of the similarities between it and the 7 pieces of cake.
  - Work through part **b** step by step.

Common mistakes and misconceptions

Turning a ratio of 4 : 5 into a fraction of $\frac{4}{5}$ is a common mistake, as is failing to find the value of the unit fraction in more complex problems such as Q5.

Plenary

- John and Jan share a lottery win in the ratio 3 : 4. Does John get $\frac{3}{4}$ of the money? If not, why not? (No, $\frac{3}{7}$)
- In total they win £7000. How much does John get? (£3000)
- John shares some of his winnings between his three children, Jo, Ben and Chang, in the ratio 1 : 3 : 2.
- What fraction does Chang get? ($\frac{2}{6} = \frac{1}{3}$)
- John gave Chang £500. How much does he give Ben? (£750)

Modular and linear specification reference

N3.1, N3.3

Keywords

denominator

Resources

Guided Practice Worksheet

7.2 Ratios and fractions

ActiveTeach Resources

Topic Tutor

Links

Follow up

Middle Practice Book 7.2

Modular and linear specification reference

N3.3

Keywords

Resources

Guided Practice Worksheet

7.3 Ratios in the form $1 : n$ or $n : 1$

ActiveTeach Resources

Links

Follow up

Middle Practice Book 7.3

Objectives

- Write a ratio in the form $1 : n$ or $n : 1$ **C**

Prior knowledge

Students should be able to carry out simple metric measurement conversions and calculate divisions with decimal answers.

Starter

Write the following ratios on the board:

a $8 : 12$ **b** $25 : 15$ **c** $2\frac{1}{2} : 3$ **d** $5 \text{ km} : 400 \text{ m}$

Ask students to simplify the ratios, then write the answers on the board and leave them there, to be use in the Main teaching.

Main teaching

- Explain that it is often easier to compare ratios when they are written in the form $1 : n$ or $n : 1$. For example, $1 : 3\frac{1}{2}$ and $1 : 2\frac{3}{5}$ are easier to compare than $2 : 7$ and $5 : 13$.

- Using the Starter questions, write all of the answers in the form:
  $1 : n$ ($1 : 1.5$, $1 : 0.6$, $1 : 1.2$, $1 : 0.08$)
  and then in the form:
  $n : 1$ ($\frac{2}{3} : 1$, $1\frac{2}{3} : 1$, $\frac{5}{6} : 1$, $12.5 : 1$).
  Show that flexibility with decimals and fractions can make things easier.

- Display Example 4 using ActiveTeach.
  - Work through parts **a** and **b** step by step.

Common mistakes and misconceptions

Students often ignore different units in a ratio and make mistakes such as simplifying 2 days : 15 hours into $1 : 7\frac{1}{2}$.

Notes on some problem-solving questions

In Q4 students can approach the problem in different ways.

- They could realise that the first ratio has 100 parts and the second ratio has 20 parts, so multiply the second by 5 and then compare the ratios.

- They could compare $78 : 22$ with $16 : 4$ by converting the ratios into the form $1 : n$. (The ratio $n : 1$ could be used, but this can tend to lead to mistakes being made.)

Plenary

Practise more of the most common exam-type questions such as Q2b–f.

Objectives

- Share a quantity in a given ratio **D**, **C**

Modular and linear specification reference

N3.3

Keywords

divide

Resources

Guided Practice Worksheet

7.4 Dividing quantities in a given ratio

ActiveTeach Resources

Animation

Links

Follow up

Middle Practice Book 7.4

Prior knowledge

Students should be able to write ratios as fractions, simplifying ratios.

Starter

Discuss sharing an amount fairly in different situations. *Raffle tickets cost £2. A and B pay £1 each. They win the £100 prize. How should they share it? What if A paid 50p and B paid £1.50?*

Main teaching

- Represent the 'shares' discussed in the Starter as ratios. Simplify ratios where possible. Emphasise that the prize should be shared in the same ratio as the initial cost.

- Discuss fair shares for the painting example. Work through Example 5 using ActiveTeach.

- Summarise the three steps for sharing in a given ratio. Leave them on display while you work through one or more examples, such as sharing £10 in the ratio 4 : 1 (or use examples from the Starter).

Common mistakes and misconceptions

For the ratio 2 : 3, some students will try to use the fraction $\frac{2}{3}$. Students usually need lots of practice to reinforce the three steps of the process.

Plenary

Suggest three or more ingredients for a fruit juice cocktail, such as orange juice, lemonade, pineapple juice, cranberry juice. Agree suitable ratios, such as twice as much orange as pineapple, and write in ratio form. *How many litres of this cocktail would you need for a class party? Work out the quantities of each ingredient.*

Objectives
- Solve word problems involving ratio **C**

Modular and linear specification reference

N3.3

Keywords

Resources

recipe (for Plenary)

Guided Practice Worksheet

ActiveTeach Resources

Topic Tutor

Links

Follow up

Middle Practice Book 7.5

Prior knowledge

Students should be able to share amounts in a given ratio, and into two or three parts.

Starter

Discuss why adult : child ratios are different for different age groups. *What is the adult : child ratio in this class?*

Main teaching

- Explain that an adult : child ratio of 1 : 3 means that for every 3 children you need 1 adult.
- There are 4 adults. How many children can they look after? Emphasise that '4 times as many adults can look after 4 times as many children'.
  - *How many adults would you need for 6 children? 9 children?* Discuss students' methods, emphasising dividing the children into groups of 3, with 1 adult for each.
- Display Example 6 using ActiveTeach and work through.
- Use the adult : child ratio of 1 : 3 again. *How many adults would you need for 10 children?* Emphasise that when you divide them into groups of 3, there is 1 left over, who needs an extra adult. Work through the calculation method, showing how the answer needs to be rounded up to give a sensible number of adults.
- Work through Example 7.

Common mistakes and misconceptions

Sometimes students don't multiply both sides of the ratio by the same number. Clear layout of working may help, as might emphasising 'doing the same to both sides' and '*n* times as many adults can look after *n* times as many children'.

Plenary

Display a recipe. *What is the ratio of flour to sugar? How much sugar do I need for n grams of flour?* Discuss why the ratio of ingredients is important.

Modular and linear specification reference

N3.3

Keywords

direct proportion, unitary method

Resources

Guided Practice Worksheet

ActiveTeach Resources

Topic Tutor

Links

Follow up

Middle Practice Book 7.6

Objectives

- Understand direct proportion **D**
- Solve proportion problems, including using the unitary method **D**

Prior knowledge

Students should be able to multiply and divide money in pounds and pence by numbers up to 12.

Starter

If I gave you £20, what would you buy? Take suggestions of items and their cost to nearest pound. Choose one, and draw up a table for cost of 1, 2, 3, 4 and 10 of these items.

Main teaching

- Using the table of costs from the Starter, illustrate that 0 items cost £0 and that as the number of items doubles, the cost doubles, etc., showing that the cost and the number of items are in direct proportion.

- Choose another item from the Starter, without telling the students what it is. Write down the cost of 5 of them. *Which item is it? How did you work it out?*

- Work through more examples. *4 CDs cost £24. How much does 1 CD cost? How much do 3 CDs cost?* Summarise the unitary method – find the cost of 1 item first.

- Display and work through Example 8 using ActiveTeach.

- Work through Example 9. Emphasise that the unitary method gives the correct answer, but using $24 = 4 \times 6$ is quicker.

Common mistakes and misconceptions

Students do not always see relationships between numbers, for example, if the cost of 4 items given, and price of 8 is asked for. Students can use the method they prefer, but should look out for these short cuts.

Plenary

Are cost and number of items always in direct proportion? What about BOGOF (buy one, get one free) and 3 for the price of 2 offers? 1 pencil costs 20p, a pack of 6 pencils costs £1.10. Are the number of pencils and cost in direct proportion?

Objectives

- Work out which product is the better buy **D**

Prior knowledge

Students should be able to divide amounts of money by 2- and 3-digit numbers, rounding answers to the nearest tenth of a penny.

Starter

Display packets of similar products, with their prices. *Which would you buy? Why?*

Main teaching

- Write down the quantity and the prices for two packets from the Starter. Explain that the better buy is the one that gives better value for money, for example, 1 g of the product at the cheaper price.
  - *How can we work out the price of 1 g in this packet?*
  - Work through the calculations and compare unit prices to find the better buy.
- Work through Example 10 using ActiveTeach. Emphasise the link with the unitary method, for finding the cost of 1 unit.
- Use the values in Example 11 to calculate cost per tissue, giving a very small answer. Explain that sometimes it is easier to work out how much you get for 1p. Work through Example 11.

Common mistakes and misconceptions

Students may get muddled and calculate price per gram for one packet, then the amount for 1p for the other. Encourage them to decide on what they are going to calculate first and write it down clearly. For example, price per gram = price ÷ quantity (g) at the top of their working.

Plenary

Discuss how some supermarkets help us decide on best buys. Labels on shelves give price per kg or per litre, for example. Is the biggest packet *always* the best buy? Consider non-financial issues. *Would you use it all before the 'use by' date?*

Modular and linear specification reference

N3.3

Keywords

best buy

Resources

packets of similar products of different sizes, with price labels

Guided Practice Worksheet

7.7 Best buys

ActiveTeach Resources

BBC Active video clip

Grade Studio: Problem solving

Links

Follow up

Middle Practice Book 7.7

Objectives
- Solve word problems involving direct and inverse proportion **D, C ,B**
- Understand inverse proportion **C, B**

Prior knowledge

Students should be able to solve direct proportion problems using the unitary method, rounding money amounts to the nearest penny and pound.

Starter

What currency do they use in other countries? Include non-euro countries. Show examples of notes and coins, if you have them. *How do you get the currency you need before you go on holiday?*

Main teaching

- Display exchange rates and explain that they vary from day to day. Display the current exchange rate for euros, such as £1 = €1.1.

- *What is £2 in euros? £10 in euros?*

- Explain that French visitors need to convert their euros into pounds. Display £1 = €1.1 again. Remind students of the unitary method – find the value of €1 first. Convert amounts in euros to pounds.

- Display and work through Example 12 using ActiveTeach.

- Remind students about direct proportion. As one value increases, the other increases at the same rate, so as one doubles, the other doubles.

- Introduce an inverse proportion scenario. *It takes one person 4 days to build a wall. How long will it take 2 people? 4 people?* Establish that this is inverse proportion – as the number of people increases, the time taken decreases. When one doubles, the other halves.

- Display and work through Example 13.

Common mistakes and misconceptions

Students often divide by the wrong quantity in conversion problems. For example, when converting $20 to pounds, encourage them to decide before they start whether the number of pounds will be bigger or smaller than the number of dollars. (It will be smaller.) This will help them check their answers. The same approach can be used for inverse proportion.

Plenary

Compare prices on multinational websites by converting both to the same currency.

Modular and linear specification reference
N3.3

Keywords
exchange rate, rounding, inverse proportion

Resources
exchange rates for a variety of currencies
coins, notes from different currencies (optional)

Guided Practice Worksheet
7.8 More proportion problems

ActiveTeach Resources
Animations (×2)
Grade Studio: Knowledge check
Topic Tutor

Links
www.exchangerate.com

Follow up
Middle Practice Book 7.8

8.1 / Repeated percentage change

Objectives
- Perform calculations involving repeated percentage changes **C**

Prior knowledge

Students should know how to find a percentage of an amount and how to convert percentages into decimals.

Starter

What is 20% of 600? (120) *What is 2% of 600?* (12)

Change these percentages to decimals:
   20%   2%   120%   102%

(Answers: 0.2, 0.02, 1.2, 1.02)

Main teaching

- *Where might you meet repeated percentage change in everyday life?* (Compound interest, prices increasing (i.e. inflation), value of goods reducing (i.e. depreciation), population changes.)
- Display Example 1 using ActiveTeach and work through.
  - Emphasise that the amount of interest earned is added on each year, so in the second year, although the percentage interest is the same, the amount is increased.
- Display Example 2. Emphasise that:
  - some situations involve a percentage decrease
  - the multiplier is expressed as a decimal
  - the multiplier is used for the number of time periods (in this case, years).

Common mistakes and misconceptions

Students may leave the multiplier as a percentage, instead of converting to a decimal. Errors are made in converting to a decimal inaccurately.

Students often fail to understand compounding (e.g. treating compound interest as simple).

Plenary

The time periods you are interested in might not be whole years. For example, the amount of a medicine in the bloodstream drops by 12% every hour. *If there is 60 mg per litre at 3 pm, how much is there by 7 pm?* (The multiplier is 0.88, so $60 \times 0.88^4 \approx 36$ mg.) The time period in this case is hours.

Modular and linear specification reference
N2.7h

Keywords
compound interest, invest, multiplier

Resources
none

Guided Practice Worksheet
8.1 Repeated percentage change

ActiveTeach Resources
Topic Tutor

Links
http://www.thisismoney.co.uk/historic-inflation-calculator

Follow up
Middle Practice Book 8.1

Modular and linear specification reference

N2.7h

Keywords

original quantity, final quantity, reverse percentage

Resources

Guided Practice Worksheet

ActiveTeach Resources

Topic Tutor

Links

Follow up

Middle Practice Book 8.2

Objectives
- Perform reverse percentage calculations **B**

Prior knowledge

Students should know how to increase or decrease quantities by a given percentage.

Starter

Use this activity to encourage students to develop the habit of estimating an answer to a percentage question. Many answers given in exams are 'silly answers' because students don't evaluate what they have written.

The price of a phone has decreased by 20%. It is now £69. Estimate its original price. (Give credit to answers between £80 and £100.) The price decreased, so it was originally more, and it is more by about one fifth.

A holiday has increased by 5% and now costs £435. Roughly what was the original cost? (Give credit to answers between £400 and £420.) The price has increased so it was originally less, and it is less by quite a small fraction (one twentieth).

Main teaching

- Explain that reverse percentages involve finding an original quantity. Emphasise that:
  - the original quantity is taken to be 100%
  - the final quantity could be less than or more than 100%.
- Display and go through Examples 3 and 4 using ActiveTeach.
  - Show that both Methods A and B give the same answer.
  - Explain that students should choose the method they are more comfortable using.

Common mistakes and misconceptions

Students may make mistakes in converting percentages to decimals (e.g. 5% = 0.5). They may fail to add an increase or subtract a decrease to the given quantity.

Plenary

The population of a town increased by 2% to 32 000. *What was the original population? Which of these is the most likely answer and why?*

38 000	31 373	32 640	26 672

Objectives

- Interpret and use standard form **B**

Prior knowledge

Students should know how to multiply and divide by factors of 10, and how to multiply and divide numbers with up to two decimal places.

Starter

What is another way of writing these numbers?

$100 \ (10^2)$ $1000 \ (10^3)$ $100 \ 000 \ (10^5)$

$0.1 \ (10^{-1})$ $0.001 \ (10^{-3})$ $1 \ (10^0)$

We can use this type of expression to help us write very large or very small numbers. For example, the mass of the Sun, as given in Why learn this?, would be 1.989×10^{30} kg.

Main teaching

- Explain why it is useful to write numbers in standard form.
  - ○ To write the mass of the Sun (for example) as an ordinary number takes up lots of space and is time consuming.
  - ○ You are more likely to make mistakes when operating with numbers with lots of zeros.
- Go through Example 5 using ActiveTeach.
  - ○ Show how to pick out the number between 1 and 10 ($1 \leq n < 10$).
  - ○ Next find what factor of 10 this number must be multiplied by to get back to the original number.
  - ○ Show how to express the factor of 10 in the form 10^x.
- Repeat the process with Example 6.
- Show the reverse process using Example 7.
- Finally, look at using calculators to answer questions involving standard form. Point out that calculators have different designs and encourage students to get to know their own calculator's features.
- Display and work through Example 8.

Common mistakes and misconceptions

Students may make careless mistakes in converting from a factor of 10 to 10^x and vice versa, and may forget to include the minus signs in the power for numbers less than 1.

Plenary

If $x = 1.2 \times 10^5$ and $y = 6.0 \times 10^3$, what is xy? What is $\frac{x}{y}$? What is $x + y$? Who can give the first correct answer for each?

Modular and linear specification reference

N1.10h

Keywords

standard form

Resources

Guided Practice Worksheet

ActiveTeach Resources

Animation
Grade Studio: Knowledge check
Grade Studio: Problem solving
Topic Tutors (×2)

Links

Follow up

Middle Practice Book 8.3

Objectives

- Multiply whole numbers using written methods **E, D**
- Use repeated subtraction for division of whole numbers **E, D**
- Round up or down in context **E, D**

Prior knowledge

Students should be able to multiply a 2-digit number by a single integer, and multiply by 10, 100 and multiples of 10 and 100.

Starter

Display these numbers:

4   24   60   14   8   2   12   72   5   16   32   36   45   9   12   56

Ask students to use the operators ×, ÷ and = to make as many true statements as possible.

Main teaching

- Display the explanatory text on the grid method and the standard method using ActiveTeach.
  - Work through the calculation of 24 × 23 using both methods. Focus on the common mistakes during the demonstration.
- Display the explanatory text for the calculation 274 ÷ 12.
  - Work through the calculation using repeated subtraction. Focus on the common mistakes during the demonstration.

Common mistakes and misconceptions

When using the grid method, students often forget to add the numbers to find the final answer. Encourage them to write 'Answer = …' before moving on. When using the standard method, students often forget the 'zero' when multiplying by tens. Encourage them to write down the calculation they are performing (e.g. 24 × 20), which should help remind them.

When dividing, students often write 3.6 when they mean 3 remainder 6. Focus on the correct notation during the lesson.

Students often forget to give an answer in the context of the problem. Ensure they check whether their answer makes sense.

Plenary

Display the calculation: 48 × 138 = 6624.

What is the value of:

    a 6624 ÷ 48?   **b** 6624 ÷ 1380?   **c** 4.8 × 138?

Modular and linear specification reference

N1.2, N1.4

Keywords

multiplication, repeated subtraction, division, divisor, remainder, round

Resources

Guided Practice Worksheet

9.1 Multiplying and dividing whole numbers

ActiveTeach Resources

Animations: Number skills (×3), 9.1 (×1)

Grade Studio: Problem solving

Topic Tutor

Links

Follow up

Middle Practice Book 9.1

Objectives

- Check and estimate answers to problems **E, D**
- Estimate answers to problems involving decimals **E, D, C**
- Make estimates and approximations of calculations **E, D, C**

Prior knowledge

Students should be able to apply the order of operations to calculations, multiply and divide by powers of 10, and round numbers to one significant figure.

Starter

Ask students to round numbers to one significant figure, such as:

a 45 286 **b** 7.3 **c** 849 **d** 6.29 **e** 0.439 **f** 496.87

Main teaching

- Display Example 1 using ActiveTeach.
  - Discuss the use of the ≈ and = symbols.
  - *What is a good approximation for 2.8 × 12.4?* Explain that the accuracy of the estimate is dependent on the context of the problem.
  - Remind students that rounding can be performed more than once during a calculation if necessary. Stress the importance of showing all their working.

Common mistakes and misconceptions

Students often find an approximate value independent of the context in which it is set. Use Q7 to illustrate the importance of considering the context. (Q7b may also highlight the common mistake of giving an answer without reading the question carefully.)

Plenary

- *Sami earns £2375 per month. Approximately how much does he earn in 1 year?*
- *A group of 23 people win £356 460 in a competition. The prize money is to be divided equally between them. Estimate how much each person will receive.*
- Display: $\dfrac{1.8 \times 584.2}{7.43 + 84.89 + 7.68}$

  Steven and Dirk use calculators to work this calculation out. Steven gets an answer of 10.52. Dirk gets 7.04. Use approximation to decide who is correct.

Modular and linear specification reference

N1.4, N1.4h

Keywords

estimate, approximate, significant figure

Resources

Guided Practice Worksheet

ActiveTeach Resources

Animations (×2)

BBC Active video clip

Topic Tutor

Links

Follow up

Middle Practice Book 9.2

Negative numbers

Modular and linear specification reference

N1.2

Keywords

positive, negative, minus sign

Resources

Guided Practice Worksheet

ActiveTeach Resources

Animation

Grade Studio: Knowledge check

Links

Follow up

Middle Practice Book 9.3

Objectives

- Multiply and divide negative numbers **E**

Prior knowledge

Students should be able to list numbers in descending/ascending order, and identify the colder/hotter temperature in a temperature pair.

Starter

Ask students to draw ten boxes vertically. Call out numbers between −20 and 20 at random without repeating any. Students must write the numbers in the boxes so that they increase in value from the bottom to the top of the ladder. If a student is unable to write a number in the correct position they are out.

Main teaching

- Review addition and subtraction of negative numbers. *Calculate:*

  $(-4) + 5$ $4 + (-6)$ $(-2) + 2$ $3 + (-7)$

  $(-3) + 9$ $9 + (-13)$ $(-4) - (+6)$ $(-4) - (-6)$

  $(-9) - (+7)$ $(-7) - (-9)$ $-11 - (-6)$ $-19 - (+3)$

- Display Example 2 using ActiveTeach.

  o As an alternative, ask students to investigate the answers to:

  2×2 2×1 2×0 $2 \times (-1)$ 2×-2

  and

  $(-2) \times 2$ (-2×1) $(-2) \times 0$ $(-2) \times (-1)$ $(-2) \times (-2)$

  o Display: $(-3) \times (-2) \times (-4)$. *The signs are the same so the answer is positive.* Discuss.

Common mistakes and misconceptions

Students often have difficulty with adding and subtracting negative numbers. Use a practical example to aid understanding: ask students to imagine adding ice to a drink (+ −) or taking ice out of a drink (− −). *What happens to the temperature of the drink?*

Students often apply the 'general rules' (*the signs are different, so the answer is negative; the signs are the same, so the answer is positive*) without constraint. Use examples to show how these 'rules' apply to calculations involving even numbers of numbers only.

Plenary

Display these numbers:

−5   −2   −6   −3   4   −25   3   21   −7   −16   −14   8   −24

Ask students to use the operators ×, ÷ and = to make as many true statements as possible.

10.1 / Multiples

Objectives
- Solve problems involving multiples **E**
- Find lowest common multiples **C**

Prior knowledge

Students should know their times tables confidently and be able to write down the next term in an arithmetic sequence.

Starter

Practise times tables. Call out numbers between 10 and 100. Ask students to write down a multiplication fact that has that number as an answer. Students could display their multiplication facts on mini-whiteboards. *Which numbers have lots of different possible multiplication facts? Which numbers only have a few?*

Main teaching

- Rehearse the 6 times table. Explain that all the answers in the 6 times table are multiples of 6.
  - *Write down the first 5 multiples of 4.* (4, 8, 12, 16, 20) Reinforce the fact that 4 is a multiple of 4.
  - *Can you write down the largest multiple of 4?* (No)
  - Make the link between multiples and tests for divisibility. *How do you know if a number is divisible by 5?* (Last digit is 5 or 0.) Explain that if a number is divisible by 5 it is a multiple of 5.
- Display Example 1 using ActiveTeach and work through.
- Ask students to write down a number which is a multiple of 4 and a multiple of 6. (e.g. 12). *Is there more than one possible answer?* (Yes) Explain that these numbers are called common multiples of 4 and 6.
- Display Example 2 to demonstrate finding lowest common multiples (LCMs) by writing out lists of multiples.
  - *Is there a largest possible common multiple?* (No)
  - Explain how the LCM can be used to solve problems. If possible demonstrate with Cuisenaire rods (e.g. *What is the shortest length I can make exactly using either 4 cm rods or 6 cm rods?*) or money (e.g. *What is the smallest amount of money I can make exactly using either 20p pieces or 50p pieces?*).

Common mistakes and misconceptions

Students often multiply numbers with a common factor when attempting to find the LCM. Encourage students to list the multiples at this stage.

Plenary

Work out the lowest common multiple of 4, 5 and 6. (60)

Modular and linear specification reference

N1.6

Keywords

multiple, common multiple, lowest common multiple (LCM), common multiple

Resources

mini-whiteboards (Starter; optional), Cuisenaire rods, money (Main teaching; optional)

Guided Practice Worksheet

10.1 Multiples
10.2 Factors and primes (combined)

ActiveTeach Resources

Links

http://Resources.oswego.org/games/Ghostblasters1/gbcd.swf
An excellent quick-fire game for recognising multiples.

http://www.mrnussbaum.com/lcmf.htm
A set of simple on-screen flashcards for finding LCMs.

Follow up

Middle Practice Book 10.1

10.2 — Factors and primes

Objectives
- Solve problems involving factors **E**
- Recognise two-digit prime numbers **E**
- Find highest common factors **C**

Modular and linear specification reference

N1.7, N1.8

Keywords

factor, prime number, prime factor, highest common factor (HCF), common factor

Prior knowledge

Students should understand multiples and be able to recognise quickly whether one number is divisible by another.

Resources

number cards or mini-whiteboards (Starter; optional)

Starter

Give pairs of students a card or miniature whiteboard with a number written on it. Call out numbers and ask pairs to stand if the number called is a multiple of their number.

Guided Practice Worksheet

10.1 Multiples

10.2 Factors and primes (combined)

Main teaching

- *What numbers is 12 divisible by?* (1, 2, 3, 4, 6, 12) Explain that these are the factors of 12.
  - o Explain that the factors of a number come in pairs, and that you can find the factors of a number by writing down all the multiplication facts with that number as an answer.
- Display Example 3 using ActiveTeach to demonstrate this method. Explain that you can check systematically for division by 2, 3, 4, etc. Reinforce that 1 and the number itself are always factors of a number.
  - o *Work out the factors of 9.* (1, 3, 9) Demonstrate that 3 occurs in a factor pair with itself ($3 \times 3 = 9$). *What type of numbers have this property?* (Square numbers)
- Explain that numbers with exactly 2 factors are called prime numbers.
  - o Explain that some of the factors of every number (except 1) will be prime numbers. These factors are called prime factors.
- Display Example 4 and work through.
- Ask students to write down a number which is a factor of 30 and a factor of 40. (1, 2, 5 or 10) *Is there more than one possible answer?* (Yes) Explain that these numbers are called common factors of 30 and 40.
- Use Example 5 to demonstrate finding HCFs by writing out lists of factors.

ActiveTeach Resources

Animations (×2)

Links

http://www.bbc.co.uk/education/ mathsfile/shockwave/games/ gridgame.html

Levels 1 and 2 of this game ask students to identify factors, multiples and primes. Level 3 requires knowledge of squares, cubes and triangular numbers.

Follow up

Middle Practice Book 10.2

Common mistakes and misconceptions

Students often forget whether they should find the highest or lowest common factor. Remind students that the *lowest* common factor of any pair of numbers is 1.

Plenary

Work out the highest common factor of 80, 96 and 100. (4)

Objectives

- Calculate squares and cubes **E**
- Calculate square roots and cube roots **E**
- Understand the difference between positive and negative square roots **D**
- Evaluate expressions involving squares, cubes and roots **C**

Prior knowledge

Students should be able to find squares of single-digit numbers and should be able to multiply with directed numbers confidently.

Starter

Complete this multiplication grid as a class:

×	−3	−7	−1	−9
−9				
−3				
−1				
−7				

Main teaching

- *What is 8 squared? (64) 12 squared? (144) 15 squared? (225) What is the square root of 81? (9)*
  - Explain that students need to know all the squares up to 15^2 and their corresponding square roots.
  - *What is $(-6)^2$? (36)* Explain that 36 has a positive square root and a negative square root.
  - Demonstrate that the cube root is unique. *What is $(-4)^3$? (-64)*
- Display Example 6 using ActiveTeach. Reinforce the need to work out any values inside a square or cube root first.

Common mistakes and misconceptions

Students often write $\sqrt{36} = -6$, or $\sqrt{36} = \pm 6$ when finding the negative square root. Reinforce the fact that the $\sqrt{}$ symbol represents the positive square root.

Plenary

What two whole numbers is $\sqrt{40}$ between? (6 and 7) Without using a calculator work out 6.5^2. Write down an estimate for $\sqrt{40}$ correct to one decimal place. Use a calculator to work out $\sqrt{40}$ correct to two decimal places.

Modular and linear specification reference

N1.7, N1.8

Keywords

square root, positive square root, negative square root, cube root

Resources

Guided Practice Worksheet

ActiveTeach Resources

Animation

Topic Tutor

Links

Follow up

Middle Practice Book 10.3

Modular and linear specification reference

N1.8, N1.9h

Keywords

index notation, index, indices, base, power

Resources

sticky notes

Guided Practice Worksheet

10.4 Indices

10.6 Index laws and standard form

(combined)

ActiveTeach Resources

Topic Tutor

Links

http://www.absorblearning.com/media/attachment.action?quick=mo&att=1624

A simple animation showing negative powers of 10.

Follow up

Middle Practice Book 10.4

Objectives

- Understand and use index notation in calculations **E**
- Understand and use negative powers and numbers to the power of 1 or 0 **B**

Prior knowledge

Students should be able to work with squares, cubes and roots confidently.

Starter

Play a reciprocals 'Pelmanism' game.

Display this grid and cover each square with a numbered sticky note. Ask students to call out pairs of numbers. Reveal these squares, then re-cover them. Students must try to match numbers with their reciprocals.

2	$\frac{1}{3}$	−0.2	0.25
$-\frac{1}{10}$	$-\frac{1}{2}$	−10	3
4	0.5	2.5	0.1
10	−5	$\frac{2}{5}$	−2

Main teaching

- Display descending positive powers of 2 and 10.

$2^4 = 16$	$10^4 = 10\,000$
$2^3 =$	$10^3 =$
$2^2 =$	$10^2 =$
$2^1 =$	$10^1 =$

  - Invite responses from the class to complete the table.

  - Ask students to identify a pattern in each column. Ask students to continue the pattern to the next step to find 2^0 and 10^0.

  - Continue the pattern to find negative powers of 2 and 10.

  - Explain that a negative power is the reciprocal of the corresponding positive power. Demonstrate this fact by pairing up the positive and negative powers in the table.

- Work through Example 7 using ActiveTeach.

Common mistakes and misconceptions

Students will often struggle to describe numbers written with powers. Rehearse common methods of describing indices (e.g. 'two to the power of four', 'six to the minus one').

Plenary

Discuss the sequence generated by $T_n = 2^n$. Ask students to generate the first six terms of this sequence. Discuss how quickly this sequence increases in size. Compare with the sequence $U_n = 1000n$. Ask students to guess the first value of n for which $T_n > U_n$. Students should check their answers using calculators ($n = 14$).

Modular and linear specification reference
N1.6

Keywords
prime factor

Resources
none

Guided Practice Worksheet
10.5 Prime factors

ActiveTeach Resources
Animation
Grade Studio: Problem solving
Topic Tutor

Links
http://nlvm.usu.edu/en/nav/
frames_asid_202_g_2_t_1.html
Interactive factor tree activity.

Follow up
Middle Practice Book 10.5

Objectives

- Write a number as a product of prime factors using index notation **C**
- Use prime factors to find HCFs and LCMs **C**

Prior knowledge

Students should understand the definitions of LCM and HCF. They should also be able to recognise prime numbers up to 100 and use multiplication facts to determine whether a number is a prime number.

Starter

Challenge students to use calculators to determine whether 3- and 4-digit numbers are prime.

Is 2035 a prime number? (No) *How do you know?* (It ends in a 5 so it must be divisible by 5.) *Is 227 a prime number?* (Yes) *How do you know?* (Try dividing by all numbers up to 16.) Discuss why there is no need to check numbers larger than 16 when checking to see if 227 is prime. (Because $16 \times 16 = 256 > 227$, so any factor > 16 will have a factor pair < 16.)

Main teaching

- Explain that you can 'break down' any number into its prime factors. All numbers can be written as a product of prime factors in exactly one way.
  - *What are the prime factors of 20?* (2 and 5) Ask students to write a calculation using 2, 5 and × with 20 as the answer (e.g. $2 \times 2 \times 5$).
- Display Example 8 using ActiveTeach and work through. Demonstrate the method of using factor trees and the method of repeated division separately.
  - Encourage students to work systematically, attempting to divide first by 2, then by 3, then by 5, etc.
- Display Example 9. Demonstrate using prime factors to find LCMs and HCFs.

Common mistakes and misconceptions

Remind students that to find an HCF they must only circle the prime factors that appear in the decompositions of *both* numbers. You can reinforce this method by writing for example,

$$60 = 2^2 \times 3^1 \times 5^1 \times 7^0 \quad \text{and} \quad 168 = 2^3 \times 3^1 \times 5^0 \times 7^1$$

and telling students to circle the lowest power of each number.

Plenary

Ask students to investigate the HCF of $2a^2b$ and $3ab^2$ for different prime numbers a and b ($a \neq b$, a, $b > 3$). *What do you notice?* (The HCF is always ab.) *Can you explain your answer? What will the LCM be?* ($6a^2b^2$)

Objectives
- Use laws of indices to multiply and divide numbers written in index notation **C, B**
- Carry out calculations with numbers given in standard form **B**

Prior knowledge

Students should be familiar with standard form from Chapter 8. They must be able to work confidently with numbers given using index notation.

Starter

Which is bigger:

   a 2^3 *or* 3^2? (3^2) **b** 3^4 *or* 4^3? (3^4) **c** 9^7 *or* 7^8? (7^8)

Students should guess and raise their hands to indicate their choice. Check answers as a class using a calculator where necessary.

Main teaching

- Revise positive and negative powers. Demonstrate the index laws for multiplication and division of numbers.

- Display Example 10 using ActiveTeach and work through. Show how the index law for division can be used to find negative powers.

- Display Example 11 and explain that you have to deal with each base separately. Explain that the index laws only apply to powers of the same number.

- Display Example 12 and revise standard form. Reinforce that the first part must be a number greater than or equal to 1 and less than 10.

  o *Which of these numbers are in standard form?*
    3×10^2 10×10^4 0.7×10^{15} 31×10^7 2.9×10^{-1}

  o Challenge students to write all the numbers from the list above in standard form.

- Work through Example 13. Explain that to add or subtract numbers in standard form you need to write them as decimal numbers. (Offer more able students the alternative strategy of writing both numbers with the same power of 10 before adding or subtracting.)

Common mistakes and misconceptions

Students often struggle to convert the answers to calculations back into standard form (e.g. $32 \times 10^{-4} \Rightarrow 3.2 \times 10^{-3}$), especially when dealing with negative powers. Make links to doubling/halving techniques of multiplication (e.g. $20 \times 18 = 10 \times 36$).

Plenary

Round 5.4659×10^{-2} to: two decimal places (0.06); one decimal place (0.1); two significant figures (0.055); three significant figures (0.0547).

Modular and linear specification reference
N1.9, N1.9h, N1.10h

Keywords
laws of indices, standard form

Resources
none

Guided Practice Worksheet
10.4 Indices
10.6 Laws of indices and standard form
(combined)

ActiveTeach Resources
Animation
Grade Studio: Knowledge check
Topic Tutor

Links
none

Follow up
Middle Practice Book 10.6

Modular and linear specification reference

N4.1, N5.1

Objectives

- Simplify algebraic expressions by collecting like terms **E**

Keywords

term, like term

Resources

cards with terms and numbers, both positive and negative, written on them, e.g. $2x$, $3y$, $-3x$, x, $+6$, -2, etc. (for Plenary)

Guided Practice Worksheet

ActiveTeach Resources

Topic Tutor

Links

Follow up

Middle Practice Book 11.1

Prior knowledge

Students should be able to simplify expressions by collecting like terms involving only one letter.

Starter

Display an addition and subtraction calculation with four terms, such as: $7 + 4 - 3 + 6$, and the same terms arranged in different order, such as: $4 + 6 - 3 + 7$ and $-3 + 6 + 7 + 4$. Ask students to calculate each, and confirm that, with addition and subtraction, order does not matter, provided the sign in front doesn't change.

Main teaching

- Display sketches of bags of marbles, one labelled x and a larger one labelled y. Tell students that the letters represent the number of marbles in each bag and that x and y are not equal to each other. *How many marbles are in both bags? ($x + y$)*
- Display three term cards, such as $4x + 2y + x$. *Does it matter which order we add these together?* (No) Demonstrate how to move the cards so that the terms in x are together: $4x + x + 2y = 5x + 2y$.
  - Some students may try to add the x and y terms together, wrongly adding $5x + 2y$ to get $7xy$. Refer students back to the bags of marbles: $5x + 2y$ means 5 of the x bags plus 2 of the y bags, and $7xy$ has no meaning.
- Display and work through Example 1 using ActiveTeach.
- Recap the meaning of perimeter and give students a couple of rectangles with given side lengths to work out the perimeter.
- Display and work through Example 2, where the side lengths are algebraic expressions.

Common mistakes and misconceptions

Students may write $3x - 2x = 1x$. Explain that it is not necessary to write the 1. Algebra is a way of saving time in describing situations such as numbers of marbles in a bag, and the convention of not writing the 1 saves more time.

Plenary

Divide the class into two teams. Shuffle the term/number cards and give three to each team. If they simplify the expression correctly they get 3 points. Give teams the option to have four cards (4 points), etc.

Modular and linear specification reference
N5.1

Keywords
algebraic

Resources
small cards with terms and numbers written on them, e.g. $2x$, $3y$, $3x$, x, 6, 2, etc. (for Plenary)

Guided Practice Worksheet
none

ActiveTeach Resources
Animation
Topic Tutor

Links
none

Follow up
Middle Practice Book 11.2

Objectives
- Multiply together two simple algebraic expressions **E**

Prior knowledge

Students should be able to multiply single-digit numbers. They should also understand multiplication as repeated addition: $n + n + n = 3$ lots of $n = 3n$.

Starter

What is 3 × 7? What is 7 × 3? What do you notice? Does this work for all multiplications? Extend to multiplications involving three numbers. For example, look at 2 × 3 × 4, asking three groups to multiply the numbers in different orders. Confirm that the order of multiplying does not affect the answer.

Main teaching

- Display the term $4y$. *What does this mean?* $(4 \times y)$ *What about $2 \times 4y$?* Show that this is 2 lots of $4y = 8y$, and also $2 \times 4 \times y = 8 \times y = 8y$. Remind students that you can multiply terms in any order.

- Practise multiplying a few more algebraic terms by numbers, such as: $3 \times 6m$, $5 \times 2p$, etc.

- Explain that just as $5m$ means $5 \times m$, xy means $x \times y$ – in algebra we do not write the multiplication sign. Give students some examples to simplify, such as: $a \times b$, $c \times d$. *What about $f \times e$?* Emphasise that letters are usually written in alphabetical order.

- Extend to square notation, such as $m \times m = m^2$ and then to terms involving numbers and letters, such as $4y \times z = 4yz$.

- Finally, demonstrate how to multiply terms with numbers and letters.

- Display the explanatory text for multiplying expressions using ActiveTeach and give students practice examples for each type of expression.

- Display Example 3 using ActiveTeach and work through.

Common mistakes and misconceptions

Students may treat terms in m^2 and in m as like terms, for example, simplifying $3m^2 + m$ wrongly to $4m^2$. Show students a square of side m and emphasise that m^2 is the area and m is a length, so the two quantities do not represent the same things and cannot be treated as like terms.

Plenary

Divide the class into two teams. Shuffle the term/number cards and give two cards to team A. If they multiply the two terms together correctly they get 2 points. Repeat with Team B with two cards initially, and then encourage teams to choose three cards (3 points), four cards, etc.

Objectives

- Multiply terms in a bracket by a number outside the bracket **D**
- Multiply terms in a bracket by a term that includes a letter **D**

Prior knowledge

Students should be able to multiply single-digit numbers by positive and negative numbers, and be able to multiply single-digit numbers by terms involving letters and numbers.

Starter

Display some rectangles with side lengths and ask students to work out the areas.

Main teaching

- Display a rectangle with sides 6 and x. *What is the area?* ($6x$) *What if we make the rectangle 4 units longer?* Display this rectangle with all the dimensions. *How can we work out the area of this new, larger rectangle?* (Either work out the two areas separately, $6 \times x = 6x$ and $6 \times 4 = 24$ and add, or work out $6(x + 4)$.) Confirm that these will give the same answer, and display $6(x + 4) = 6x + 24$.

  o Demonstrate how to multiply each term in the bracket by the term outside, and that this gives the same answer: $6x + 24$.

  o *Does this always work?* Test with some numbers, such as: $5(2 + 7)$ – work out $5 \times 9 = 45$, and $5 \times 2 + 5 \times 7 = 45$ to confirm they give the same answer.

- Display and work through Example 4 using ActiveTeach. Advise students to look out for minus signs.

- Recap multiplying letter terms such as $m \times m$, $m \times n$, $2m \times m$, $3m \times 2n$.

- Work through Example 5.

Common mistakes and misconceptions

Students forget to multiply the second term in the bracket by the term outside, expanding $2(x + 3)$ incorrectly as $2x + 3$, or they ignore minus signs, writing $3(m - 2)$ as $3m + 6$.

Plenary

Display two equivalent expressions with one term missing, such as: $6(m + 2)$ and $6m +$ ☐. *What goes in the box?*

Try also expressions such as $4(n +$ ☐ $)$ and $4n + 12$, to introduce informally the idea of factorising.

Modular and linear specification reference

N5.1

Keywords

brackets, expand

Resources

Guided Practice Worksheet

11.3 Expanding brackets

ActiveTeach Resources

Animations (×3)

Topic Tutor

Links

Follow up

Middle Practice Book 11.3

11.4 Simplifying expressions with brackets

Objectives
- Simplify expressions involving brackets **D**, **C**

Prior knowledge

Students should be able to expand single brackets with positive and negative terms outside the bracket, and simplify expressions, collecting like terms.

Starter

Ask some quick-fire times table questions, then extend to negatives, such as: 2×-3, -4×5, -3×-5.

Main teaching

- Display the expression $4(x + 3)$ and ask students to expand it. Repeat for $3(2x + 5)$. Leave the two expansions clearly on display.

- Display the expression $4(x + 3) + 3(2x + 5)$.
  - *How can we simplify this algebraic expression?* Expand the brackets, then substitute the expansions, to give $4x + 12 + 6x + 15$.
  - *What kinds of term do we have in this expression?* (Numbers and terms in x.) Ask students to simplify the expression by collecting like terms: $10x + 27$.

- Display and work through the two parts of Example 6, reminding students to be careful where there are minus signs.

- Give extra practice examples with one negative value either inside or outside one of the brackets, for example, $3(m - 1) + 2(m + 5)$, $4(n + 2) - 3(n + 1)$.

Common mistakes and misconceptions

Students forget to multiply the second term in the bracket by the term outside, expanding $2(x + 3)$ incorrectly as $2x + 3$, or they get the wrong signs when multiplying negative values. Encourage them to write out each step clearly and decide on the sign before multiplying the two values each time.

Plenary

Divide the class into teams. Shuffle the expressions cards. Give each team two cards. Spin a coin – heads means subtract one expression from the other, tails means add. Correct simplifications gain 2 points. Collect in the cards and shuffle again.

After a few rounds, teams can choose to have three expressions cards, to gain 3 points.

Modular and linear specification reference

N5.1

Keywords

Resources

set of around 20 cards, each with an expression involving brackets, e.g. $3(x + 4)$, $5(2x - 1)$, etc., coin (for Plenary)

Guided Practice Worksheet

ActiveTeach Resources

Topic Tutor

Links

Follow up

Middle Practice Book 11.4

Objectives

- Recognise factors of algebraic terms **D**
- Simplify algebraic expressions by taking out common factors **D**

Prior knowledge

Students should know factors of numbers, and common factors of pairs of numbers.

Starter

Recap the meaning of factor. Emphasise that if 3 is a factor of a number, you can write that number as $3 \times \square$.
Give students a set of numbers such as 15, 6, 8, 27, 12, 10, 13.
Which have 3 as a factor? Write the $3 \times \square$ multiplication for each one.

Main teaching

- *What does 3m mean? ($3 \times m$) We can write 3m as $3 \times \square$, so 3 must be a factor of 3m. We can write 3m as $\square \times m$, so m must be a factor of 3m. What are the two factors of 7p? (7 and p)*
- Display and work through Examples 7 and 8 using ActiveTeach.
- Display the expression $3x + 15$.
  - *I gave an expression with brackets in to Jim. He expanded it, and this is his result. What expression did I give him in the first place?*
  - Display: $\square\,(\square + \square)$. *What number could have gone outside the bracket?* Encourage students to give the factors of $3x$ and 15, and find the common factor, 3.
  - *Now what terms were inside the bracket?* Consider $3 \times \square = 3x$ and $3 \times \square = 15$, to give $3(x + 5)$.
  - Explain that factorising is the opposite or inverse of expanding.
- Display and work through Example 9.
- Remind students to start each factorisation by writing a common factor of both terms outside the brackets. Work through Example 10.

Common mistakes and misconceptions

Students identify the common factor, but forget to work out one of the terms inside the bracket. Sometimes they don't realise that x is a factor of x and x^2.

Plenary

Play a factorising/expanding memory game, with a set of 20 cards containing 10 expressions with brackets and their expansions. Spread the cards face down, then take turns to pick two. If the two expressions are equivalent, the player keeps them and takes another turn. If they are not, they replace them and play passes to the next player. The winner is the player with most pairs when all the cards have been picked up.

Modular and linear specification reference

N5.1

Keywords

factor, common factor, factorise

Resources

set of 20 cards: 10 showing expressions involving brackets, 10 showing expansions of those expressions (for Plenary)

Guided Practice Worksheet

11.5 Factorising algebraic expressions

ActiveTeach Resources

Animation

Topic Tutor

Links

Follow up

Middle Practice Book 11.5

> ## Objectives
> - Multiply together two algebraic expressions with brackets **C**, **B**
> - Square a linear expression **C**, **B**

Prior knowledge

Students should be able to add and subtract like terms, and be able to multiply letters and numbers.

Starter

Use the grid method to multiply two 2-digit numbers, such as 34×25:

×	30	4
20	600	80
5	150	20

$34 \times 25 = 600 + 150 + 80 + 20 = 850$

Main teaching

- Display a square with sides x. *What is its area?* (x^2)
  - Add 2 units to the top side and 5 units to the left side to give a rectangle $x + 2$ by $x + 5$. *How could you work out the area of this rectangle?* $(x + 2) \times (x + 5)$.
- Display Example 11 using ActiveTeach. Remind students that in algebra we don't need to write the multiplication sign.
  - Work out the areas of the four smaller rectangles as shown in Example 11, and add them together to give
    $(x + 5)(x + 2) = x^2 + 7x + 10$.
- Display Example 12 to demonstrate the grid method for multiplying two brackets.
- Use Example 12 to demonstrate the FOIL method of expanding two brackets (multiply **F**irst, **O**uter, **I**nner, **L**ast). Confirm that this gives the same result for the multiplication as the grid method.
- Work through Example 13 using either the grid method or FOIL to multiply two brackets with non-unit x coefficients.

Common mistakes and misconceptions

Students may forget to multiply pairs of terms; remind them that FOIL has four letters, and should give them four terms to work with.

Plenary

Repeat Exercise 11K Q6 for a grid with five numbers in each row. *What is the final answer each time?* (5) *What about a grid with seven numbers in each row?* Test students' predictions.

Modular and linear specification reference

N5.1h

Keywords

Resources

Guided Practice Worksheet

11.6 Expanding two brackets

ActiveTeach Resources

Animation

Grade Studio: Knowledge check

Grade Studio: Problem solving

Topic Tutor

Links

Follow up

Middle Practice Book 11.6

Comparing fractions

Objectives

- Compare fractions with different denominators **E**, **D**

Prior knowledge

Students should be able to divide one whole number by another to give a decimal answer, and be able to find lowest common multiples.

Starter

Give students some equivalent fraction questions such as:

a $\frac{1}{4} = \frac{\square}{12}$ **b** $\frac{2}{5} = \frac{\square}{15}$

Ask students to tell you three or four equivalent fractions to 35.

Ask students for a common denominator for the following sets of fractions:

c $\frac{1}{2}$ and $\frac{3}{5}$ **d** $\frac{2}{3}$ and $\frac{5}{7}$ **e** $\frac{1}{2}$, $\frac{2}{5}$ and $\frac{9}{20}$

Main teaching

- Explain to students that comparing and ordering fractions without a calculator is a common exam question. It can either be done by converting each fraction into a decimal by division, or more easily, by converting to fractions with a common denominator.

- Display Example 1 using ActiveTeach.
  - Show the link between the example and Q2b in the Skills check.

Common mistakes and misconceptions

Students may multiply the denominator but not the numerator when finding equivalent fractions in order to compare.

Plenary

Ask students to put the fractions $\frac{3}{4}$, $\frac{4}{5}$ and $\frac{7}{10}$ in order.
Which of the three fractions is nearest to $\frac{19}{25}$?

Modular and linear specification reference

N1.5, N2.1

Keywords

common denominator

Resources

Guided Practice Worksheet

12.1 Comparing fractions

ActiveTeach Resources

Grade Studio: Problem solving

Topic Tutor

Links

Follow up

Middle Practice Book 12.1

Objectives
- Add and subtract fractions when one denominator is a multiple of the other **E**
- Add and subtract fractions when both denominators have to be changed **D**

Prior knowledge

Students should be able to add and subtract simple fractions, and find common multiples.

Starter

Go over Q1a in Exercise12A again, then ask students to work out $\frac{3}{4} - \frac{5}{7}$.

Do the same with Q1b and ask students to work out $\frac{7}{8} - \frac{5}{6}$.

Do the same for Q1c and Q1d, asking the students to work out:

a $\frac{4}{9} - \frac{2}{5}$ **b** $\frac{3}{7} - \frac{5}{12}$.

Main teaching

- Display Example 2 using ActiveTeach.
  - Work through both parts, referring back to the Starter where possible.

Common mistakes and misconceptions

Students may multiply the denominator but not the numerator when finding equivalent fractions.

Plenary

Write these fractions on the board:

$$\frac{1}{2} \quad \frac{2}{3} \quad \frac{3}{4} \quad \frac{2}{5} \quad \frac{1}{6}$$

Tell the students they are competing against each other. You are going to point to any two fractions and say either 'sum' or 'difference'. Once they have worked out the answer to each question, they must write their answer on their mini-whiteboard and hold the board in the air.

Modular and linear specification reference

N2.2

Keywords

denominator, equivalent fractions, numerator

Resources

mini-whiteboards (for Plenary)

Guided Practice Worksheet

ActiveTeach Resources

Animations (×2)

Topic Tutor

Links

Follow up

Middle Practice Book 12.2

Modular and linear specification reference

N2.2

Keywords

mixed number, improper fraction, lowest terms

Resources

Guided Practice Worksheet

12.3 Adding and subtracting mixed numbers

ActiveTeach Resources

Topic Tutor

Links

Follow up

Middle Practice Book 12.3

Objectives

- Add and subtract mixed numbers **C**

Prior knowledge

Students should be able to convert mixed numbers into improper fractions, and convert improper fractions into mixed numbers.

Starter

Ask students to work out the following calculations:

a $\frac{2}{3} - \frac{1}{5}$ **b** $\frac{2}{9} + \frac{3}{4}$ **c** $\frac{5}{6} - \frac{1}{4}$

Ask students to describe in words an improper fraction and a mixed number.

Main teaching

- Emphasise that when adding and subtracting improper fractions, the method is exactly the same as adding and subtracting the fractions in the Starter.
- Display Example 3 using ActiveTeach.
  - Work through both parts, emphasising their similarity to the Starter.

Common mistakes and misconceptions

Students may multiply the denominator but not the numerator when finding equivalent fractions.

Plenary

Put students into small groups and ask them what possible answers they could get if they were asked to work out $1\frac{\square}{4} + 2\frac{\square}{3}$. Discuss with students what values they used for the boxes. Did anyone use 0, or numbers greater than 3 or 4?

Objectives
- Multiply a fraction by a fraction **E**

Prior knowledge

Students should be able to convert mixed numbers into improper fractions, and convert improper fractions into mixed numbers. They should also be able to find a fraction of an amount, and find a fraction of a fraction.

Starter

Give students some easy fraction × number questions to do:

a $\frac{1}{3}$ of 18   **b** $\frac{2}{3} \times 12$ **c** $\frac{4}{5} \times 20$

Show students how to set them out correctly, for example:

$$\frac{1}{3} \text{ of } 18 = \frac{1}{3} \times 18 = \frac{1}{3} \times \frac{18}{1} = \frac{18}{3} = 6$$

and also how to cancel before multiplying, for example:

$$\frac{2}{3} \times 12 = \frac{2}{3} \times \frac{12}{1} = \frac{2}{1\cancel{3}} \times \frac{\cancel{12}^4}{1} = 8$$

Main teaching

- Display Example 4 using ActiveTeach.
  - Work through each part step by step, pointing out the similarities between these and the Starter.

Common mistakes and misconceptions

When multiplying a fraction by a fraction, students may multiply diagonally as though 'cross-multiplying' is being done, for example: $\frac{2}{3} \times \frac{5}{6} = \frac{12}{15}$

Plenary

Write these fractions on the board:

$$\frac{1}{2} \quad \frac{2}{3} \quad \frac{3}{4} \quad \frac{2}{5} \quad \frac{1}{6}$$

Tell the students they are competing against each other. You are going to point to any two fractions and ask them to multiply them together. Once they have worked out the answer to each question, they must write their answer on their mini-whiteboard and hold the board in the air.

Modular and linear specification reference

N1.2

Keywords

cancel

Resources

mini-whiteboards (for Plenary)

Guided Practice Worksheet

ActiveTeach Resources

Animations (×2)

Topic Tutor

Links

Follow up

Middle Practice Book 12.4

Objectives

- Multiply a whole number by a mixed number **D**
- Multiply a fraction by a mixed number **C**
- Multiply a mixed number by a mixed number **B**

Modular and linear specification reference

N2.7

Keywords

Resources

Guided Practice Worksheet

ActiveTeach Resources

Animation

Links

Follow up

Middle Practice Book 12.5

Prior knowledge

Students should be able to convert mixed numbers into improper fractions, and convert improper fractions into mixed numbers. They should also be able to find a fraction of an amount, and find a fraction of a fraction.

Starter

Give students some fraction × fraction questions to do, such as:

a $\frac{2}{5} \times \frac{10}{12}$ **b** $\frac{1}{3} \times \frac{7}{8}$ **c** $\frac{3}{1} \times \frac{5}{6}$

Remind students to cancel as much as possible before multiplying.

Main teaching

- Display Example 5 using ActiveTeach.
  - Work through each part step by step, pointing out the similarities between the parts and the Starter.

Common mistakes and misconceptions

When multiplying a whole number by a fraction, students may multiply both the numerator and the denominator by the whole number, such as:

$3 \times \frac{5}{6} = \frac{15}{18}$

Plenary

Ask students to work out the answers to these questions:

a $3 \times 2\frac{1}{5}$ **b** $\frac{3}{4} \times 2\frac{2}{5}$ **c** $3\frac{3}{4} \times 2\frac{3}{5}$

Objectives
- Find the reciprocal of a whole number, a decimal or a fraction **C**

Modular and linear specification reference

N1.3

Keywords

reciprocal

Resources

Guided Practice Worksheet

12.6 Reciprocals

ActiveTeach Resources

Links

Follow up

Middle Practice Book 12.6

Prior knowledge

Students should know how to divide 1 by whole numbers and decimals, and be able to convert improper fractions into mixed numbers.

Starter

Ask students to work out the answers to these questions:

$$\mathbf{a}\ \frac{1}{5} \times \frac{5}{1} \qquad\qquad \mathbf{b}\ \frac{2}{5} \times \frac{5}{2} \qquad\qquad \mathbf{c}\ \frac{3}{5} \times \frac{5}{3}$$

Ask students what they notice about each question and each answer. Introduce the word reciprocal.

Main teaching

- Read through the explanatory text on finding reciprocals using ActiveTeach, linking the first and second points to the Starter.
- Display Example 6 using ActiveTeach.
  - Parts **a** and **c** are the easier parts to do as they are already in whole number and fraction form. Part **b** needs more work, as it is a decimal. Make sure students understand the method for finding the reciprocals of decimals.

Common mistakes and misconceptions

Students may leave denominators as decimal numbers, and may not simplify answers when asked to do so.

Plenary

Check that students can work out and explain to each other how to work out the reciprocals of 2, 0.2 and $2\frac{1}{2}$.

12.7 Dividing fractions

Objectives

- Divide a whole number or a fraction by a fraction **D**
- Divide mixed numbers or fractions by whole numbers **C**
- Divide mixed numbers by mixed numbers **B**

Prior knowledge

Students should be able to find common factors and cancel fractions to their simplest form. They should also be able to multiply fractions by fractions or mixed numbers.

Starter

Practise fraction multiplication questions by asking students to work out the answers to these calculations:

a $10 \times \frac{2}{3}$ **b** $\frac{2}{7} \times \frac{3}{4}$ **c** $3\frac{1}{2} \times 7$ **d** $2\frac{1}{2} \times 2\frac{3}{5}$

Main teaching

- Display Example 7 using ActiveTeach.
  - Work through each part step by step ensuring that students understand the method.
- Display Example 8.
  - Work through each part step by step, ensuring that students understand that the method is the same as Example 7, except that the mixed numbers need to be changed into improper fractions first.
- If students want to know why the method of 'inverse and multiply' works, then you could explain using the following example:

$$\frac{\frac{4}{7}}{\frac{3}{5}} = \frac{4}{7} \times \frac{1}{\frac{3}{5}}$$

Explain that $\frac{1}{\frac{3}{5}}$ is the equivalent of the reciprocal of $\frac{3}{5}$, which is $\frac{5}{3}$.

So, $\frac{4}{7} \times \frac{1}{\frac{3}{5}} = \frac{4}{7} \times \frac{5}{3} = \frac{20}{21}$

Common mistakes and misconceptions

Students may find the reciprocal of the wrong fraction, and then multiply, or find the reciprocal of both fractions, and then multiply.

Plenary

Ask students to explain their methods for questions in Exercise 12G.

Modular and linear specification reference

N1.2

Keywords

Resources

Guided Practice Worksheet

ActiveTeach Resources

Animation

Grade Studio: Knowledge check

Topic Tutor

Links

Follow up

Middle Practice Book 12.7

Objectives

- Add and subtract decimal numbers **E**

Modular and linear specification reference

N1.1, N1.2

Keywords

decimal number

Resources

Guided Practice Worksheet

13.1 Adding and subtracting decimals

ActiveTeach Resources

Links

Follow up

Middle Practice Book 13.1

Prior knowledge

Students should be able to add and subtract whole numbers.

Starter

Ask students to do a few additions and subtractions of whole numbers and troubleshoot any problems.

Main teaching

- Make sure students know that the decimal point comes at the end of a whole number, and that the decimal point always separates the whole number from the fraction part.
- Display Example 1 using ActiveTeach.
  - Explain that adding and subtracting decimals follows exactly the same principles as adding whole numbers. Numbers must be aligned so that the units column and all the other columns match up, with the decimal points underneath each other.
  - Go through part **a**. A decimal point is added to the end of any whole numbers, and when adding, add down a column, ignoring any blank spaces.
  - Go through part **b**, stressing that once the numbers have been written correctly aligned, it is just like adding and subtracting whole numbers.

Common mistakes and misconceptions

If students do not write their numbers carefully enough in clear columns, they may become confused about which digits are in the same column when they come to add/subtract.

Plenary

Ask a student to do a subtraction such as 30 – 0.03. Look at the formal written method and compare with any other ways the students have of doing the calculation.

Modular and linear specification reference

N2.3

Keywords

Resources

Guided Practice Worksheet

13.2 Converting decimals to fractions

ActiveTeach Resources

Links

Follow up

Middle Practice Book 13.2

Objectives

- Convert decimals to fractions **D**

Prior knowledge

Students should be able to cancel fractions. Ideally, they should be able to add and subtract fractions, but as a minimum, they should have some appreciation of the problem of adding fractions if the denominators are different.

Starter

Get the class to cancel some simple fractions, then include some fractions like $\frac{10}{100}$ and $\frac{10}{1000}$, thus highlighting the relationships between the column headings for decimal numbers.

Main teaching

- Display Example 2 using ActiveTeach.
  - o Go through part **a**. Explain that for 0.7, students just have to work out which column the 7 is in, and then write that as a fraction.
  - o For 0.48, encourage them to locate the right-hand column, $\frac{1}{100}$s in this case, and then make the decimal part the numerator, to get the fraction $\frac{48}{100}$.
  - o Then stress that students must always try to cancel. If you think it will help you can explain that they are always looking for common factors of 2 and/or 5 when cancelling these fractions.
  - o Go through part **b**. Explain that they can put the whole number part 'on one side' and concentrate on the decimal part, but must remember to add the whole number back on at the end.
  - o It is probably worth going through the cancelling for this example in case some students are put off by the size of the numbers initially involved.

Common mistakes and misconceptions

If there are several decimal places, students can get the power of 10 wrong. Students can be confused by zeros in the middle of a number, as in Exercise 13B Q1 and Q2.

Plenary

Write some amounts of money on the board and ask the students to convert these to fractions. Stress that it is useful to know that 75p is $\frac{3}{4}$ of £1, etc.

13.3 Multiplying and dividing decimals

Objectives
- Multiply and divide decimal numbers **D**, **C**

Prior knowledge

Students should be able to multiply and divide using whole numbers and know that any fraction can be thought of as 'numerator ÷ denominator'.

Starter

Give the students a few long multiplications to do and troubleshoot any problems they have.

Main teaching

- Go through the long multiplication 125×73, using the method the students are used to using.
- Display Example 3 using ActiveTeach.
  - Explain that if $1.25 = 125 \div 100$ and $7.3 = 73 \div 10$, then the answer will need to be divided by 1000.
  - Show this leads to the rule that you can count the decimal places in the numbers in the multiplication to place the decimal point in the answer. Do another example on the whiteboard to ensure method is understood.
- Display Example 4.
  - Explain that dividing a decimal by a whole number works exactly like a normal division, with the decimal points aligned vertically.
- Display Example 5.
  - Explain that, having written the division as a fraction, we must multiply by 10, 100, etc., whichever number is big enough to make the denominator a whole number.
  - Then do the division on the whiteboard if you deem it helpful.

Common mistakes and misconceptions

Students may confuse multiplication with the rules for addition, writing a long multiplication with the decimal points underneath each other.

In a decimal division, students may try to 'move the decimal point back' at the end.

Plenary

Challenge students to find two numbers that add up to 5 and multiply together to give the largest possible total. (Answer: $2.5 \times 2.5 = 6.25$.) Extend to other numbers – is there a rule?

Modular and linear specification reference
N1.2

Keywords
none

Resources
none

Guided Practice Worksheet
13.3 Multiplying and dividing decimals

ActiveTeach Resources
Animations (×2)
Grade Studio: Problem solving
Topic Tutors (×2)

Links
none

Follow up
Middle Practice Book 13.3

13.4 Converting fractions to decimals

Modular and linear specification reference

N2.3, N2.4

Keywords
terminate, recurring

Resources
none

Guided Practice Worksheet
13.4 Converting fractions to decimals

ActiveTeach Resources
Grade Studio: Knowledge check
Topic Tutor

Links
none

Follow up
Middle Practice Book 13.4

Objectives

- Convert fractions to decimals **D**
- Recognise recurring decimals **C**
- Understand how recurring decimals relate to fractions **B**

Prior knowledge

Students should know a fraction is 'numerator ÷ denominator', and be able to cancel fractions.

Starter

Ask students to sketch three circles, shading $\frac{1}{3}$ of one, $\frac{2}{5}$ of the second and $\frac{3}{7}$ of the third. Discuss the difficulty of putting them in order of size with any certainty. Being able to convert fractions to decimals will offer a solution to this problem. (It might be helpful to have an answer ready to display on the whiteboard.)

Main teaching

- Display Example 6 using ActiveTeach.
  - Go through the parts, explaining the term terminating decimal.
- On the whiteboard go through the division converting $\frac{1}{3}$ to a decimal.
  - Explain the term recurring decimal, and show how we write this with a dot over the 3 to indicate a recurring digit: $\dot{3}$.
  - Show how to indicate recurring decimals where two or more digits recur.
  - Display Example 7.
  - Go through the example explaining that we can stop as soon as we know we have a recurring digit.

Common mistakes and misconceptions

Students may confuse 0.3 with $\frac{1}{3}$.

Students can find it hard to understand this is exact maths, and mistakenly round their answers.

Plenary

Return to the fractions of the Starter and put them in order of size, possibly adding one or two others (e.g. any of $\frac{3}{8}, \frac{4}{11}, \frac{7}{20}, \frac{9}{25}$).

14.1 Solving two-step equations

Objectives

- Solve two-step equations like $2x - 1 = 11$ **E**, **D**

Prior knowledge

Students should be able to solve one-step equations using the balance method. They should also be able to work with fractions and mixed numbers.

Starter

Discuss how two-step expressions (such as $2x + 1$, $3x - 1$, $\frac{x}{5} - 1$, $\frac{x-5}{2}$) might be constructed. Use function machines if this helps. Discuss how to 'undo' the expressions using inverse operations in each case. Mini-whiteboards would be useful in this for this activity.

Main teaching

- Explain that equations can be thought of as balanced scales. Display the explanatory text using ActiveTeach.

  o *How can you find out how many marbles there are in each bag?*

  o Work through the example using the balanced scales analogy. Explain the final step carefully – students will need to understand that if there are a third of the marbles on the left-hand side (LHS) of the scale there will need to be a third of the marbles on the right-hand side (RHS).

- Display Example 1 using ActiveTeach.

  o Ask students to explain how the expression on the LHS of the equation has been constructed. Again, use function machines if this helps.

  o Ask students how they would 'undo' the expression and then use inverse operations to solve the equation. Remind students that an equation is like a set of balanced scales and inverse operations need to be applied to both sides of the equation.

- Display and explain Examples 2 and 3.

Common mistakes and misconceptions

Students may have difficulty with equations that are written in a different format. For example, $7 + 2a = 9$ and $9 = 2a + 7$ may cause problems where $2a + 7 = 9$ does not. Students may also find it difficult to combine number work involving fractions and decimals with solving equations.

Plenary

Ensure all students can check answers using substitution.

Modular and linear specification reference

N5.4

Keywords

Resources

mini-whiteboards (for Starter)

Guided Practice Worksheets

14.1a–14.4a Solving equations 1
14.1b–14.4b Solving equations 2

ActiveTeach Resources

Animation
Topic Tutor

Links

http://www.mangahigh.com/
http://www.bowlandmaths.org.uk/
http://mathsnet.net/algebra/

Follow up

Middle Practice Book 14.1

Objectives
- Write and solve equations **E, D**

Prior knowledge

Students should be able to solve one- and two-step equations using the balance method. They should also be able to work with fractions and mixed numbers.

Starter

Work as a class and then in pairs to play 'I think of a number'. Ask questions such as: *I think of a number, multiply it by 2 and then add 3. My answer is 7. What is the number I was thinking of?*

Main teaching

- Explain that equations can be useful to solve problems in the real world but it is necessary to construct them first. Display Example 4 using ActiveTeach.
  - Explain that the expression represents the length 14 cm so the two things can be linked using an equals sign.
  - Ask students to solve the equation $2y + 4 = 14$ when it has been constructed.

Common mistakes and misconceptions

Students may have difficulty with equations that are written in a different format. For example, $7 + 2a = 9$ and $9 = 2a + 7$ may cause problems where $2a + 7 = 9$ does not. Students may also find it difficult to combine number work involving fractions and decimals with solving equations.

Plenary

Ask students to work in pairs and write their own questions involving measuring rods similar to Example 4 for their partner to solve.

Modular and linear specification reference
N5.4

Keywords
none

Resources
none

Guided Practice Worksheets
14.1a–14.4a Solving equations 1
14.1b–14.4b Solving equations 2

ActiveTeach Resources
Topic Tutor

Links
http://www.mangahigh.com/
http://www.bowlandmaths.org.uk/
http://www.math.com/school/subject2/

Follow up
Middle Practice Book 14.2

Objectives
- Solve equations involving brackets **D**, **C**

Prior knowledge

Students should be able to solve one- and two-step equations using the balance method. They should also be able to expand brackets and simplify algebraic expressions.

Starter

Students can expand algebraic brackets using mini-whiteboards. Include brackets with negative coefficients and unknowns.

Main teaching

- Explain that when solving equations with brackets it is necessary to expand the brackets first.
- Display Example 5 using ActiveTeach.
  - Work through the example.
  - *What does the equation look like when the brackets on the left-hand side have been expanded?*
  - Ask students to solve the equation $8a - 4 = 12$ and then go through the rest of the example ensuring students have solved it correctly.

Common mistakes and misconceptions

Mistakes are most commonly made when expanding brackets involving negative numbers.

Students can sometimes struggle when presented with more than one set of brackets or if some further simplification is required after expanding the bracket.

Plenary

Discuss other ways of solving equations involving brackets. For example: $4(2a - 1) = 12$ can also be solved by dividing both sides by 4. The equation becomes $2a - 1 = 3$.

Ask students to solve some equations from Exercise 14E in a similar way and compare their answers.

Modular and linear specification reference
N5.4

Keywords
expand

Resources
none

Guided Practice Worksheets
14.1a–14.4a Solving equations 1
14.1b–14.4b Solving equations 2

ActiveTeach Resources
Animation
Topic Tutor

Links
http://www.mangahigh.com/
http://www.bowlandmaths.org.uk/
http://www.math.com/school/subject2/
http://mathsnet.net/algebra/l2_equation.html

Follow up
Middle Practice Book 14.3

Objectives

- Solve equations with an unknown on both sides **D, C**

Prior knowledge

Students should be able to solve one- and two-step equations using the balance method. They should also be able to expand brackets and simplify algebraic expressions.

Starter

Ask students to represent the following equations as sets of balanced scales using a bag of marbles to represent the unknown:

a $2x + 1 = x + 4$ **b** $5x + 2 = 3x + 8$ **c** $4x = 2x + 6$

Ask students to construct their own equations with an unknown on either side and represent the equations using balanced scales. Discuss how to change the scales so there are bags of marbles on only one side. Remind students that they need to maintain the balance.

Main teaching

- Display Example 6 using ActiveTeach.
  - Explain that when the unknown appears on both sides, the best first step is usually to change the equation to one where the unknown only appears once. In this case $4x$ can be subtracted from both sides leaving $3x - 12 = 3$.
  - Ask students to solve the equation $3x - 12 = 3$ and then go through the rest of the example ensuring students have solved it correctly.

- Display Example 7.
  - Explain that sometimes it is best to add unknowns, then subtract them so that you end up with a positive number of unknowns. The best step here is to add $2x$ to both sides of the equation leaving $4 = 5x - 6$ which students can solve.

Common mistakes and misconceptions

Mistakes are most commonly made when there are a negative number of unknowns on either side of the equation. Student may struggle with equations written in formats that are unfamiliar, so $7 = 9 - 2x$ would probably cause more problems than $-2x + 9 = 7$.

Plenary

Construct and solve hot cross questions (Exercise 14G Q4).

Modular and linear specification reference

N5.4

Keywords

Resources

mini-whiteboards

Guided Practice Worksheets

14.1a–14.4a Solving equations 1
14.1b–14.4b Solving equations 2

ActiveTeach Resources

Topic Tutor

Links

http://www.mangahigh.com/
http://www.bowlandmaths.org.uk/
http://mathsnet.net/algebra/l2_equation.html

Follow up

Middle Practice Book 14.4

14.5 Equations with fractions

Objectives
- Solve equations involving fractions **C**, **B**

Prior knowledge

Students should be able to solve equations using the balance method. They should also be able to expand brackets and simplify algebraic expressions, and be familiar with lowest common multiples (LCMs).

Starter

Find the LCMs of numbers on mini-whiteboards.

Main teaching

- Display Example 8 part **a** using ActiveTeach.
  - Explain that when a fraction has a denominator of 4, the numerator is being divided by 4. This can be eliminated by using inverse operations: multiply by 4.
  - Work through the first part of the example and then ask students to solve $3x + 10 = 28$.
  - Display Example 8 parts **b** and **c** and tell students that multiple denominators can be eliminated by multiplying the equation by the LCM.
  - Multiply the equation in part **b** by 12 and show students how to add the fractions (or alternatively cancel).
  - Ask students to simplify and solve the remaining equation.
  - Show students how to multiply by 6 in part **c**. Ask students to simplify and solve the remaining equation.

Common mistakes and misconceptions

Mistakes are made most when cancelling after multiplying by the LCM.

Plenary

Play 'I think of a number' ensuring that students use equations involving division or fractions and ensure students can check their answers using substitution.

Modular and linear specification reference

N5.4

Keywords

eliminate, denominator

Resources

mini-whiteboards (for Starter)

Guided Practice Worksheet

ActiveTeach Resources

Links

http://www.mangahigh.com/

http://www.bowlandmaths.org.uk/

http://www.mathsteacher.com.au/

Follow up

Middle Practice Book 14.5

Objectives

- Represent inequalities on a number line **E**
- Write down whole-number values for unknowns in an inequality **D**
- Solve inequalities **C, B**

Prior knowledge

Students should be able to to draw number lines and solve inequalities using the number line.

Starter

Show pairs of values including negatives and decimals and ask students to raise their left or right hand depending on which number is bigger, or use mini-whiteboards to solve equations using the balance method.

Main teaching

- Display Example 9 using ActiveTeach.
  - Show how to represent inequalities on number lines.
- Display Example 10.
  - Explain that inequalities can be solved using the balance method. In part **a**, show how all parts of the inequality can be divided by 3 to isolate x.
  - Ask students to represent the result on a number line.
  - Work through part **b**.
- Tell students that the direction of the sign needs to be changed if they are multiplying or dividing by a negative number.
- Display Example 11.
  - Work through both parts using students' experience of solving similar equations.

Common mistakes and misconceptions

Students forget to reverse the sign when multiplying or dividing by a negative.

Plenary

Investigate what happens when both sides of an inequality are multiplied or divided by a negative number.

Use $x > 5$: *What happens when you multiply by 2? Is the inequality still true? What about adding 2? Subtracting 2? Dividing by 2? What about multiplying or dividing by –2? What about adding or subtracting –2?*

Modular and linear specification reference

N5.7, N5.7h

Keywords

inequality

Resources

mini-whiteboards (for Starter)

Guided Practice Worksheet

ActiveTeach Resources

Topic Tutors (×2)

Links

http://www.mathsrevision.net/gcse/

Follow up

Middle Practice Book 14.6

Modular and linear specification reference
N5.4h

Objectives

- Solve a pair of simultaneous equations **B**

Prior knowledge

Students should be able to simplify expressions, solve linear equations, and use substitutions.

Starter

Play 'Substitution bingo' – ask students to draw a 4 by 5 grid and fill it with the numbers 1 to 20 in any order. Write $a = 2$, $b = 3$ and $c = 5$ on the whiteboard and call out expressions using combinations of a, b and c with answers between 1 and 20. The first person to get a line wins.

Main teaching

- Explain that simultaneous equations have two unknowns. Methods used to solve them involve using algebraic steps to eliminate one of the unknowns before attempting to solve the remaining equation.
- Display Example 12 using ActiveTeach.
  - Work through part **a** explaining that the method works because the equations contain the same number of ys.
  - *How is part **b** different to part **a**?*
  - Students should see that there is a negative y ($-y$) and a positive y ($+y$) instead of two positive ys. *What would happen if we subtracted equation 2 from equation 1 in part **b**?* Write $-y - +y$ on the whiteboard. Work through the rest of part **b**.
- Display Example 13.
  - *How is Example 12 different to Example 13?*
  - Students should see that the pair of equations have different numbers of unknowns so cannot simply be added or subtracted.
  - Work through the example.
- Display Example 14 and work through it.

Common mistakes and misconceptions

Students make mistakes when deciding whether to add or subtract equations.

Plenary

Ensure students can check their answers using substitution.

Keywords
simultaneous equation, eliminate

Resources
mini-whiteboards (for Starter)

Guided Practice Worksheet
none

ActiveTeach Resources
Animation
Grade Studio: Knowledge check
Grade Studio: Problem solving
Topic Tutor

Links
http://www.mathsrevision.net/gcse/

Follow up
Middle Practice Book 14.7

Objectives

- Use index notation in algebra **E**, **D**, **C**
- Use index notation when multiplying or dividing algebraic terms **D**, **C**

Prior knowledge

Students should be able to find squares and cubes using the square and cube keys on a calculator.

Starter

Ask some quick-fire questions on squares of numbers up to 10 and cubes of 1–5 and 10.

Main teaching

- Using examples from the Starter, emphasise that, for example, $6^2 = 6 \times 6$ and $2^3 = 2 \times 2 \times 2$, and that we read 6^2 as '6 squared' and 2^3 as '2 cubed'. Display $a \times a$. How could we write this in a shorter way? (a^2) Introduce the keywords index and indices.

- Repeat for $a \times a \times a$. Then extend to higher powers, such as 4 and 5. Emphasise that we read these as 'a raised to the power 4', and so on. Explain that a is a variable – it can take a variety of different values.

- *How do we write $3 \times b$ in algebra?* ($3b$) *What about $3 \times b \times b$?* ($3b^2$) Demonstrate how to simplify $2b \times 4b$ by writing it as $2 \times b \times 4 \times b$ and rearranging. ($2 \times 4 \times b \times b = 8b^2$)

- Display Example 1 using ActiveTeach and work through.
  - Demonstrate multiplying powers of the same letter, such as $b^2 \times b^3$, by writing them out in full first ($b \times b \times b \times b \times b$) and then writing in index form. (b^5)

- Display Example 2 and work though part **a**. Extend to terms including numbers, such as $3e^2 \times 4e$, and work through Example 2 part **b**.
  - Demonstrate dividing powers of the same letter, such as $m^5 \div m^3$, writing them in full and cancelling. Work through Example 2 part **c**. Extend to terms involving numbers and indices, such as $8n^4 \div 4n^3$.

Common mistakes and misconceptions

Students do not realise that x means x^1, or that a number divided by 1 = the number, for example $6 \div 1 = 6$.

Notes on some problem-solving questions

In Exercise 15B Q3 and Q6, point out that a quicker way of multiplying is to add the powers, and of dividing is to subtract the powers. Illustrate with examples from preceding questions.

Plenary

Divide the class into teams. A team picks two cards showing terms in x. Spin a coin to decide whether they multiply or divide them. Give points for correct answers.

Modular and linear specification reference

N1.9

Keywords

squared, cubed, power, index, variable

Resources

set of cards with terms in x and powers of x, e.g. x, x^2, x^3, $2x$, $3x$, $3x^2$, $4x^3$, etc., coin (for Plenary)

Guided Practice Worksheet

ActiveTeach Resources

Topic Tutor

Links

Follow up

Middle Practice Book 15.1

15.2 Laws of indices (index laws)

Objectives
- Use index laws to multiply and divide powers in algebra **C**
- Raise a number or variable to the power of 1 or 0 **C**, **B**
- Use index laws for raising a power to another power **B**

Modular and linear specification reference

N1.9

Keywords

index law

Resources

set of cards (optional, for Plenary)

Guided Practice Worksheet

15.2 Laws of indices (index laws)

ActiveTeach Resources

Topic Tutors (×2)

Links

Follow up

Middle Practice Book 15.2

Prior knowledge

Students should be able to write multiplications of variables and numbers using index notation.

Starter

Display multiplications involving powers of letters and numbers, and ask students to simplify them, such as: $5^2 \times 5^3 = 5^5$, $a^2 \times a^2 = a^4$.

Main teaching

- Using examples from the Starter, highlight the powers in the question and the answer. *What is the link?* Establish that the powers are added.

- Repeat for divisions, for example $6^4 \div 6^3$, setting out on two lines and cancelling. Highlight the powers in the question and the answer. *What is the link?* Establish that the powers are subtracted.

- Display the index laws for multiplication and division and give students plenty of practice with simple examples with letters or numbers only.

- *How can you simplify $(m^2)^3$?* Write it as $m^2 \times m^2 \times m^2 = m^6$. Repeat for more examples with letters and powers of numbers. *What is the link between the powers?* Display the index law for raising to a power.

- Display: $2^2 = 4$
  $2^3 = 8$
  $2^4 = 16$

  o Point out that going down the board you multiply by 2 each time. Going up the board you divide by 2 each time.

  o Extend to 2^0. Repeat for other numbers. Encourage students to notice that $x^1 = x$ and $x^0 = 1$ for all x.

- Display Example 3 using ActiveTeach and work through.

- Display Example 4 and work through.

Common mistakes and misconceptions

Students get confused with unit and zero powers, stating that x means x^0, or that $x^0 = 0$.

Plenary

Give students 'missing number' questions, such as:

$n^4 \times n^\square = n^7$ $3m^2 \times \square m = 12m^3$ $x^\square \div x^3 = x^5$ $a^\square = 1$ $(b^2)^\square = b^6$

You may wish to prepare these on cards in advance, and run the session as a quiz.

Modular and linear specification reference

N4.2

Keywords

formula

Resources

Guided Practice Worksheet

ActiveTeach Resources

Topic Tutor

Links

Follow up

Middle Practice Book 15.3

Objectives

- Use algebra to write formulae in different situations **E**, **D**

Prior knowledge

Students should be able to write algebraic expressions to represent real-life situations.

Starter

Ask students what their favourite sweets are, and how much they cost. *How much would 5 cost? How much would 10 cost? How did you work it out?*

Main teaching

- Write formulae for the cost in pence of different numbers of packets of the sweets suggested in the Starter. First write word formulae, then change to letters, for example:

  cost = price in pence × number of packets $C = 45 \times n$

  o *Who would buy those sweets?* Use the formula to work out the cost for that number of students. Remind them to convert the answer in pence to pounds and pence.

  o Explain that a formula must contain an equals sign, and the letters in a formula are called variables – they can take different values.

- Ask each student to choose one type of sweet from the class list and record the choices. Use the formulae to work out the total cost for the class, for example: $C = 45n + 30m + 15x$

- *How much does a DVD cost? Write a formula to work out the cost of different numbers of these DVDs.* Emphasise that price of DVD is in pounds, so total cost is in pounds.

- Display Example 5 using ActiveTeach and work through.

  o Give students similar examples to do on their own, for example: Jay buys n packets of sweets. Each packet costs 30p. Write a formula for the change he gets from £10.

Common mistakes and misconceptions

Students often find it difficult to see the 'general' case. Encourage them to think how to work out the cost of 2, 5, 10 items first, and to write their formula in words before 'converting' it into algebra.

Plenary

Display the formula $M = 5n + 10$. *What could this formula be for?* (Total cost of n tickets at £5 each plus £10 booking fee; time to cook something.) Emphasise that you need to state what the letters stand for in any formula you write.

Ask students to write their own formulae using letters and ask the class to guess what they could be for.

Objectives

- Substitute numbers to work out the value of simple algebraic expressions **E**
- Substitute numbers into expressions involving brackets and powers **D, C**

Prior knowledge

Students should know the order of operations, including brackets, indices and be able to calculate with negative numbers.

Starter

Show a bag of counters; do not say how many counters are in the bag. Agree on a letter to represent the number of counters, such as x. *What if I add 2 more counters?* ($x + 2$) Write expressions for other situations, such as removing 3 counters, adding an identical bag. Display the expressions.

Main teaching

- Give a student the bag from the Starter and ask them to count the counters inside. Go back to the expressions and show that, now that the value of x is known, you can write this value instead of x in all the expressions and evaluate them. Confirm that replacing letters in an expression with numbers is called substituting.

- Display the expression: $4a$. Tell them that $a = 5$. *What is $4a$ now?* ($4 \times 5 = 20$) You may need to remind students that $4a$ means $4 \times a$. Extend to products of two letters, such as xy, $3cd$.

- Recap the rules for calculating with negative numbers.

- Display Example 6 using ActiveTeach and work through.

- Remind students of BIDMAS, emphasising evaluating brackets and then indices first. Give them a few number expressions to evaluate, such as $5(4 + 3)$ and $3^2 - 2(3 + 1)$.

- Ask a few quick-fire questions on squares of positive and negative numbers up to 10, and positive square roots.

- Display Example 7 and work through.

Common mistakes and misconceptions

When substituting $a = 6$ into the expression $4a$, students write 46 and assume it is forty-six. Emphasise writing the expression with multiplication signs first, before substituting.

Plenary

Display a set of expressions involving the four operations and brackets and indices (such as a selection from the exercises in this section). Choose an expression, then pick cards from a set of −10 to +10 number cards to generate the values to substitute for the letters.

Modular and linear specification reference

N4.2, N5.6

Keywords

substitute, evaluate

Resources

bag of counters (for Starter)

set of −10 to + 10 number cards (for Plenary)

Guided Practice Worksheet

ActiveTeach Resources

Animation

Topic Tutor

Links

Follow up

Middle Practice Book 15.4

15.5 Substituting into formulae

Objectives
- Substitute numbers into a variety of formulae **E**, **D**

Prior knowledge

Students should know the order of operations, including brackets and indices and square roots.

Starter

Recap solving two-step equations, such as $2x + 4 = 10$ and $5y - 7 = 3$.

Main teaching

- Display a rectangle, with length l and width w labelled. *How do we work out the perimeter of this rectangle?* As a class, derive the formula: $P = 2l + 2w$. Repeat for the area, $A = lw$.

- Display more rectangles with values for length and width marked. Demonstrate how to substitute these values into formulae they have just derived to calculate perimeter and area of each.

- Display Example 8 using ActiveTeach, which uses formulae involving division, a fraction and powers.

- Display Example 9 and work through, explaining how substitution results in an equation to solve. For more practice, use the area and perimeter formulae again: *When P = 48 and l = 15, what is w?* (9)

Common mistakes and misconceptions

When substituting $a = 6$ into the expression $4a$, students write 46 and assume it is forty-six. Emphasise writing the expression with multiplication signs first, before substituting.
Students do not realise that $\frac{n}{10}$ means $n \div 10$, or that $\frac{1}{2} \times 6$ means $\frac{1}{2}$ of $6 = 3$.

Plenary

Display a 'joke' formula, such as:

teacher's happiness quotient $= \frac{n - m - p}{R}$

where n = number of questions students complete in class, m = number of times teacher has to tell class to be quiet, R = number of students in class and p = number of days till Friday. Agree values for the variables and evaluate.

Repeat for students' own formulae.

Modular and linear specification reference
N4.2, N5.6

Keywords
none

Resources
none

Guided Practice Worksheet
15.5 Substituting into formulae

ActiveTeach Resources
Animation
BBC Active video clips (×2)
Topic Tutor

Links
http://news.bbc.co.uk/1/hi/health/2630869.stm
For frequently quoted happiness formula.

Follow up
Middle Practice Book 15.5

Objectives
- Rearrange a formula to make a different variable the subject of the formula **C, B**

Modular and linear specification reference
N5.6

Keywords
subject, rearrange

Resources
completed cards from GPW 15.6 (for Plenary)

Guided Practice Worksheet
15.6 Changing the subject of a formula

ActiveTeach Resources
Animation
Grade Studio: Knowledge check
Grade Studio: Problem solving
Topic Tutor

Links
none

Follow up
Middle Practice Book 15.6

Prior knowledge

Students should be able to solve one- and two-step equations, using the balance method.

Starter

Display the formula: $F = 15h + 25$. Tell students that Tom is an electrician. He uses this formula to work out his fee for a job, where h is the number of hours, and F is the fee in pounds. One week Tom charges these fees for four different jobs: £55, £70, £100, £40. *How many hours did he spend on each job?* (2, 3, 5, 1)

Main teaching

- Ask students to explain how they solved the problem in the Starter. Establish that they substituted each value for F and solved the equation to find h, so they solved four equations in total.

- *Is there a way to do this more quickly? What value in the formula do we want to find?* (h) Emphasise that we want a statement $h = ...$ Demonstrate how to rearrange the formula to make h the subject, using the same techniques as for rearranging and solving equations. Use the rearranged formula to work out h when $F = $ £175 and £115. ($h = 10$ and 6)

- Display Example 10 using ActiveTeach, emphasising that the letter on its own on one side of the equals sign is the subject of the formula.

- Display Example 11 and work through to show how to change the subject of more complex formulae including square roots and powers.

Common mistakes and misconceptions

Many students find rearranging formulae very difficult. They may find it easier to represent formulae in function machines and draw the inverse function machines to rearrange. For example, rewriting $c = 2a + 5$ as:

$a \rightarrow \boxed{\times 2} \rightarrow \boxed{+ 5} \rightarrow c$ so the inverse is:

$a \leftarrow \boxed{\div 2} \leftarrow \boxed{- 5} \leftarrow c$ rewritten as:

$\dfrac{c - 5}{2} = a$

Plenary

Use the completed cards in GPW 15.6 to play 'pairs' as a class or in small groups. Students spread out the cards face down, and take turns to take two cards. If the cards are rearrangements of each other, they keep them and have another go. If not, they replace them and play passes to the next player. The winner has most pairs when all the cards have been picked up.

Percentage increase and decrease

Objectives
* Calculate a percentage increase or decrease **D**

Prior knowledge

Students should be able to work out a percentage of an amount.

Starter

Ask students to think of examples of percentage increases and decreases that they meet in everyday life.

Main teaching

* Two methods for working out percentage increases and decreases, labelled Methods A and B, are described in the explanatory text. Method A is a little more straightforward, but Method B becomes increasingly important in later work. If the group is very weak it is suggested you focus mainly on Method A, but any who can should be encouraged to understand and use Method B as well.

* Display Example 1 using ActiveTeach.
  o Explain the methods by going through the example. Stress that Method A is just like previous work – students just need to add the amount on as a final step.
  o Encourage students to set their work out clearly. Stress that it really helps to prevent mistakes.

* Display Example 2.
  o Explain the methods. Stress the similarity with percentage increase, but with the single difference in both methods that the addition becomes a subtraction.

Common mistakes and misconceptions

With Method A students can forget to do the final step.

Students can also get into bad habits, writing '=' between quantities that are not equal, because they use the sign as a shorthand for 'then I do this'.

Plenary

A real example of a price increase or discount from a recent newspaper or local shop would make a good example to round off the lesson. Students could then be asked to bring their own example from real life to the next lesson.

Modular and linear specification reference
N2.6, N2.7

Keywords
original amount, percentage increase, percentage decrease, discount, reduce

Resources
none

Guided Practice Worksheet
16.1 Percentage increase and decrease

ActiveTeach Resources
Animations: Number skills (×2)
Topic Tutor

Links
none

Follow up
Middle Practice Book 16.1

Modular and linear specification reference

N2.5, N2.7

Keywords

VAT, credit, hire purchase, deposit, interest, interest rate, simple interest, per annum

Resources

Guided Practice Worksheets

16.2a Calculations with money: VAT and interest

16.2b Calculations with money: Credit

ActiveTeach Resources

Topic Tutor

Links

Follow up

Middle Practice Book 16.2

Objectives

- Perform calculations involving VAT **D**
- Perform calculations involving credit **E**
- Perform simple interest calculations **E**

Prior knowledge

Students should be able to calculate a percentage increase.

Starter

Ask students what they know about VAT. Discuss examples such as VAT on till receipts or on invoices for goods or services.

Main teaching

- Explain the basic principles of VAT. Explain that some goods have no VAT, such as books, most leisure activities, and most food. Some items pay a reduced VAT rate, such as children's car seats and domestic fuel.

- Display Example 3 using ActiveTeach. Stress that this is just an example of a percentage increase, so the maths is not new to them.

- Explain the basic idea of how credit works. Explain that the main advantage is that the buyer does not have to find the full purchase price straightaway, but the disadvantage is that the total amount paid in the end is usually more. Take care to explain the terminology and language. *It is very important to understand how to calculate the total cost of buying on credit before agreeing to any purchase.*

- Display Example 4. Explain that the deposit is often expressed as a percentage of the cash price. This is a typical example showing the sort of difference between the total credit and cash prices.

- Ask the class what they understand about how banks and building societies work. Give a brief explanation that banks lend at a higher rate than they pay savers, and this is their source of income.

- Display Example 5. Explain that the interest for 1 year is a simple percentage calculation. If the interest is taken out, then the amount in the account stays the same, so if the interest rate stays the same the interest for 5 years is just 5 times the interest for one year.

Common mistakes and misconceptions

In VAT calculations students can trip up over percentages that are not whole numbers, such as 17.5%. In credit calculations, they can forget to add on the initial deposit.

Plenary

Show an example of a recent advert for a sofa or a car, and work out how much more you pay on credit.

Modular and linear specification reference

N2.7

Objectives
- Calculate a percentage profit or loss **C**

Prior knowledge

Students should be able to work out one quantity as a percentage of another.

Starter

Ask the class to do a question involving finding one quantity as a percentage of another.

Write a few fractions, such as $\frac{2.5}{7.5}$ and $\frac{4.5}{18}$, on the whiteboard and discuss with the students the best way to simplify them.

Main teaching

- Explain the terms cost price and selling price.
- Display Example 6 using ActiveTeach.
  - Explain that the profit is the difference between the cost and selling prices, and go through the example. The Starter should have helped them see how to do the calculation fairly simply.
  - Remind the students that they are basically finding one quantity as a percentage of another.
- Display Example 7.
  - Explain the term depreciation, and explain that this is just like the previous example except that this time there is a loss, which is the cost price less the selling price.

Common mistakes and misconceptions

Some students can get confused between cost price and selling price if the question is complex or long.

Plenary

The percentage mark-up on groceries is anything from 5% to 25%. The mark-up on shoes is 40% or more. The mark-up on jewellery is usually 100% or more. Put up a couple of examples for the class to do, finding the percentage profit, and discuss why there are such differences.

Keywords
cost price, selling price, profit, loss, percentage profit, percentage loss, depreciation

Resources
none

Guided Practice Worksheet
16.3 Percentage profit or loss

ActiveTeach Resources
none

Links
none

Follow up
Middle Practice Book 16.3

16.4 Repeated percentage change

Objectives
- Perform calculations involving repeated percentage changes **D**

Modular and linear specification reference
N2.6, N2.7

Keywords
compound interest, multiplier

Resources
none

Guided Practice Worksheet
16.4 Repeated percentage change

ActiveTeach Resources
Topic Tutor

Links
none

Follow up
Middle Practice Book 16.4

Prior knowledge

Students should be able to calculate a percentage increase or decrease. By this stage students really need to be able to cope with calculating a percentage change using a multiplier.

Starter

Give students an example of a percentage increase and an example of a percentage decrease to calculate.

Main teaching

- Explain that usually when money is put into a savings account the interest is just added to the amount in the account, so the amount in the account increases. As the amount in the account grows, so the amount of interest grows.

- You could explain how banks lend at a higher rate of interest than they pay to savers, and that is how they finance their business.

- Display Example 8 using ActiveTeach. It is important to stress that the interest changes each year because the amount in the account has grown, in contrast to simple interest.

  o Show how the calculations can be simplified using a multiplier.

- Display Example 9. In this case the bike loses 10% of its value each year.

Common mistakes and misconceptions

Students can find these questions difficult because they are less familiar with the multiplier method, and don't fully understand it. They can get confused about when the multiplier should be greater than or less than 1.

Plenary

Bring to the class some current interest rates for savings accounts and compare how savings would grow in different accounts. After one or two manual calculations you could display results on a spreadsheet, so showing results for several years.

Objectives
- Perform calculations involving finding the original quantity **B**

Prior knowledge

Students should be able to calculate a percentage change. It is desirable but not essential that they can calculate a percentage change using a multiplier.

Starter

Give students an example of a percentage increase and an example of a percentage decrease to work out.

Main teaching

- Reverse the percentage decrease question in the Starter so that it is now an example of a question finding the original quantity (to which they know the answer). Invite the students to say how they would solve the problem, pretending they did not know the answer. The most common wrong way of answering the problem (just treating it as a percentage increase question) should arise in the discussion.
- Display the explanatory text for Methods A and B using ActiveTeach.
- Display Example 10, and go through the example. Choose the method more suitable for the abilities of the group.
- Display Example 11, showing that reversing a percentage increase can be handled in the same way.

Common mistakes and misconceptions

The most common mistake is to fail to realise that the problem is not a straightforward percentage increase/decrease question.

Notes on some problem-solving questions

In Exercise 16G Q10, encourage students to calculate the correct answer first. Then consider how John got his wrong answer, and the explanation should follow.

Plenary

Put up a couple of questions involving finding the original quantity and a couple which are straightforward percentage increase/decrease questions. Discuss with the students which are which, how to recognise them (do we have the 100% amount?), and the method for solving them.

Modular and linear specification reference
N2.6, N2.7h

Keywords
reverse percentage, original quantity

Resources
none

Guided Practice Worksheet
16.5 Reverse percentages

ActiveTeach Resources
Grade Studio: Knowledge check
Grade Studio: Problem solving
Topic Tutor

Links
none

Follow up
Middle Practice Book 16.5

17.1 Number patterns

Modular and linear specification reference

N6.1

Keywords

sequence, term, consecutive, term-to-term rule

Resources

Guided Practice Worksheet

ActiveTeach Resources

Topic Tutor

Links

Follow up

Middle Practice Book 17.1

Objectives

- Find the next term in a sequence **E, D**
- Describe the rule for continuing a sequence **E**

Prior knowledge

Students should be familiar with the ordinal numbers, and be able to subtract numbers giving solutions less than zero.

Starter

Write the sequence 3, 6, 9, 12, … and ask students to decide whether 18, 21, 20 and 35 will be in the sequence. Discuss why.

Main teaching

- Display Example 1 using ActiveTeach and work through.
- Write the first five square numbers on the board: 1, 4, 9, 16, 25
  - Ask students to find the difference between consecutive numbers.
  - Show the differences on the board in the following way:

1		4		9		16		25
	$\rightarrow$		$\rightarrow$		$\rightarrow$		$\rightarrow$	
	+ 3		+ 5		+ 7		+ 9	

  - *What should we add to find the next term in the sequence?* (11)
  - *What is the next term in the sequence?* (36)
  - *Is this a square number?* (Yes)
- Display Example 2 and work through.

Common mistakes and misconceptions

Students may expect all sequences to have common differences. Stress this is not the case.

Plenary

Write the sequence of triangular numbers on the board: 1, 3, 6, 10, 15, 21, …

Ask students to find:

a the next term in the sequence   **b** the 10th term

Discuss with students how they found these terms. Discuss the limitations of finding the next few terms in a sequence by looking at previous terms.

Objectives

- Find any term of a sequence given a formula for the nth term **E**
- Find the nth term of a linear sequence **C**

Prior knowledge

Students should be able to substitute positive integers into algebraic expressions (both linear and involving small powers of the unknown).

Starter

Write the following algebraic expressions on the board and ask students to find their value for different values of x (use positive integer values):

a $4x$ **b** $x - 5$ **c** $\frac{x}{3}$ **d** $2x + 4$

e x^2 **f** $3x^2$ **g** $x^2 - 8$

Main teaching

- Discuss with students the limitations of finding the next few terms of a sequence by considering differences.
- Explain that sequences can be defined by a general term.
  - o Display $n - 5$ and describe it as the nth or general term.
  - o To find any term in the sequence substitute the term number for n.
  - o Show students how to find the 1st, 2nd, 3rd and 4th terms of the sequence inviting class to contribute. Repeat for more terms
  - o Discus the advantages of the general term over looking at differences. Continue the sequence round the class.
- Display Example 3 using ActiveTeach and work through step by step
- Discuss the relationship between the differences between consecutive terms and the general term. Establish that the general term of a sequence can be found by considering differences.
- Display Example 4 and work through.

Common mistakes and misconceptions

Students may mistake x^2 for $2x$. Ask students to try both for values of x.

Plenary

How do we find the general term of a linear sequence? (By comparing the sequence to multiples of the difference between consecutive terms.)
What is the general term of this sequence: –5, 1, 7, 13, 19? ($6n - 11$)
Why is it useful to be able to find the general term? (We can then find any term in the sequence without working out all the terms in between.)

Modular and linear specification reference

N6.2

Keywords

nth term, general term, position-to-term rule, linear sequence, multiple

Resources

Guided Practice Worksheets

17.2a Rules for sequences: Using differences
17.2b Rules for sequences

ActiveTeach Resources

Topic Tutor

Links

Follow up

Middle Practice Book 17.2

Modular and linear specification reference

N6.2

Keywords

Resources

Guided Practice Worksheet

ActiveTeach Resources

Topic Tutors (×2)

Links

Follow up

Middle Practice Book 17.3

Objectives

- Find the *n*th term of a linear sequence **C**
- Use the *n*th term to find terms in a sequence **C**

Prior knowledge

Students should be able to find any term in a sequence given the general term.

Starter

Display these numbers:

  2   3   6   10   15   16   17   20   25

Ask students (working in pairs) to decide which terms will be in the sequence generated by the general term:

  a $2n$ **b** $3n$ **c** $5n$

Main teaching

- Review with students how to find the *n*th term of a linear sequence.
- Display the first part of Example 5 using ActiveTeach and find the *n*th term of the sequence.
  - Hence find the 50th term.
  - Discuss with students how to identify if a term is in the sequence. Elicit that they must solve the equation.
  - Work through part **b** of the example.

Common mistakes and misconceptions

Students think the only way to find the 50th term is to find all the terms up to that term. Encourage them to use the *n*th term.

Plenary

Write the *n*th term of a sequence on the board: $2n + 1$

Write the following numbers on the board and ask students if they are terms in the sequence or not:

  2   5   8   20   32   39   200   343

What can you say about all the terms in the sequence? (They are odd.)

What about the terms in the sequence:

  a $2n - 1$ (odd)   **b** $2n + 2$ (even)   **c** $3n + 1$ (odd or even)

17.4 Sequences of patterns

Objectives

- Find the next few terms in a sequence of patterns **E**
- Find the *n*th term for a sequence of diagrams **C**

Prior knowledge

Students should be able to find the *n*th term of a linear sequence.

Starter

Display the sequence: $\frac{1}{2}$ $\frac{2}{3}$ $\frac{3}{4}$

- *What are the next three terms in the sequence?* ($\frac{4}{5}, \frac{5}{6}, \frac{6}{7}$)
- *Is the sequence ascending or descending?* (Ascending)
- *What is this sequence of numbers getting closer to?* (1)
- *What is the 10th term?* ($\frac{10}{11}$) *What is the 20th term?* ($\frac{20}{21}$)
- *How is the numerator of the fraction related to the denominator?* (The denominator is 1 more.)
- *What is the nth term of the sequence?* ($\frac{n}{n+1}$)

Main teaching

- Draw the following on the board:

  - o Ask students to come to the board and draw the next two terms.

  - o *How many extra dots are there in each term of the sequence?* (2)
    How many pairs of dots are there in the 1st term, 2nd term, 3rd term?
    What else appears in each term? (A single dot)
    How many pairs of dots will there be in the 10th term? (10)
    How many dots in total will there be in the 10th term? (21)
    How many dots will there be in the nth term? ($2n + 1$)

- Display Example 6 using ActiveTeach and work through step by step.

Common mistakes and misconceptions

Some students may not initially make the connection between the structure of the physical pattern and the form the *n*th term takes.

Plenary

Refer students to the sequence of triangular numbers (Q2). Ask students to work out the differences between terms. *Is this a linear sequence?* (Are the differences constant?) Explain that some sequences do not have common differences but have a pattern in the differences. Elicit from the students that the pattern of differences is increasing whole numbers.

Modular and linear specification reference

N 6.2

Keywords

pattern

Resources

Guided Practice Worksheet

ActiveTeach Resources

Topic Tutor

Links

www.mathsisfun.com/pascals-triangle.html

Follow up

Middle Practice Book 17.4

Modular and linear specification reference

N6.1

Keywords

constant, quadratic

Resources

Guided Practice Worksheet

ActiveTeach Resources

Links

Follow up

Middle Practice Book 17.5

Objectives

- Find the first few terms of a quadratic sequence by using the nth term **D**
- Find the next few terms of a quadratic sequence by looking at differences **D**

Prior knowledge

Students should be able to find the nth term of a linear sequence, and be able to substitute values into a quadratic expression.

Starter

Draw a sequence of squares with sides of length 1, 2, 3, 4, … on the board.

Ask students to write down the sequence of perimeters and areas of the squares. Find the general term for the perimeter ($4n$) and the area (n^2).

Main teaching

- Display the sequence 3, 9, 15, 21, …
  - ○ *What is the next term?* (27) *How did you find the next term?* (By considering differences.)
  - ○ Explain that this is a linear sequence because the differences are constant. *What is the general term of the sequence?* ($6n - 3$) Explain that all linear sequences will have a general term of this form.
- Display Example 7 using ActiveTeach and work through step by step.

Common mistakes and misconceptions

Students may mistake x^2 for $2x$. Ask students to try both for values of x.

Plenary

Draw the following sequence on the board:

Ask a student to come to the board and draw the third term in the sequence.

How many dots will there be in the 5th term? (30)
How did you work it out? (5×6)
Draw the following table on the board:

Term number	1	2	3	4	5	n
Number of dots	1×2	2×3				

Ask individual students to fill in the missing values. [3×4, 4×5, 5×6, $n \times (n + 1)$]

Ask students to expand the general term giving $n^2 + n$.

Modular and linear specification reference

N6.1

Keywords

Resources

Guided Practice Worksheet

ActiveTeach Resources

Links

Follow up

Middle Practice Book 17.6

Objectives

- Find the *n*th term of a simple quadratic sequence **C**

Prior knowledge

Students should be able to find the *n*th term of a linear sequence, and be able to find any term in a quadratic sequence given the general term.

Starter

Ask students to find the 1st, 2nd, 3rd and 10th terms in the sequence generated by:

a $2n + 5$ **b** n^2 **c** $3n^2$ **d** $n^2 - 12$

Main teaching

- Write the sequence 0, 3, 8, 15, 24, … on the board.

  o *What is the next term?* (35)
  How did you find the next term? (By considering differences.)

  o Ask students to look at the differences and then the second differences.

  o Explain that if the second difference is constant, then the sequence is a quadratic sequence and the general term will be related to n^2.

- Ask students to give the first five terms of the sequence generated by the general term n^2.

  o *How does our sequence relate to the sequence?* (It is 1 less.)

  o *What is the general term of our sequence?* ($n^2 - 1$)

- Display Example 8 using ActiveTeach and work through step by step.

Common mistakes and misconceptions

Students may think $(3x)^2 = 3x^2$. Remind them of the order of operations.

Plenary

Ask students what the first 10 cubed numbers are.

Look at the differences until they are constant. (The third row of difference.)

Repeat for the first 10 numbers raised to the power 4. Ask students to predict how many rows of differences they will have to consider before they are constant.

Objectives
- Show step-by-step deduction when proving results **E, D, C**

Modular and linear specification reference

N5.9

Keywords

verify, proof

Resources

Guided Practice Worksheet

17.7 Proof

ActiveTeach Resources

Grade Studio: Problem solving

Links

Follow up

Middle Practice Book 17.7

Prior knowledge

Students should be able to use letters to represent unknowns.

Starter

Write the following list of numbers on the board:

17   31   32   85   100

Split students into pairs and ask them to decide which of the list of numbers are terms in the sequence $3n - 5$. (31, 85, 100)

Discuss with students their methods for finding the terms. (Trial and improvement; solving equations; recognising that all the terms must be 5 less (or 2 less) than a multiple of 3.)

Main teaching

- Write on the board: 'All students in the class are over 1 m tall.'
  - Discuss with students how they would prove a statement like this.
  - Explain that it is not enough to simply show that a few students are over 1 m tall. All students must be over 1 m for the statement to be true.
- Write on the board: 'All fish have gills.'
  - Discuss how they could verify this (by checking several fish).
  - Discuss how they would prove it (check all fish, living and dead).
  - Explain that in mathematics in order to prove something is true we must show it is true for all cases.
- Display and work through Examples 9 and 10 using ActiveTeach.

Common mistakes and misconceptions

Students must understand that proof is not showing something works for a few values – it is showing it is true for all values.

Plenary

Display: 'The sum of two consecutive triangular numbers = a square number.' Ask a student to come and draw a diagram associated with the triangular numbers.

Show how to make them into square numbers in the following way:

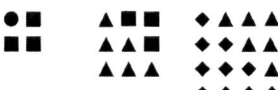

Explain that a proof can be visual.

Modular and linear specification reference

N5.9

Keywords

counter-example

Resources

Guided Practice Worksheet

17.8 Using counter-examples

ActiveTeach Resources

Grade Studio: Knowledge check

Links

Follow up

Middle Practice Book 17.8

Objectives

- Show something is false by using a counter-example **C**

Prior knowledge

Students should be able to use letters to represent unknown variables and understand the difference between proof and verification.

Starter

Write on the board: 'x is an integer'. Explain that knowing this you can state many more facts, such as '$2x$ is an even number'.

Split students into pairs and ask them to write down as many other facts (algebraically) as they can, such as:

$3x$ is a multiple of 3	$2x - 1$ is an odd number
$(2x)^2$ is an even number	$10x$ is a multiple of 10
$x(x + 1)$ is an even number	$x(x - 1)$ is an even number

Main teaching

- Remind students that to prove something you must show it is true in all cases – it is not sufficient just to give a couple of examples.

- Explain that disproof is sometimes easier. Write this statement on the board: 'All maths teachers play squash.'

  o *How might you disprove this?* (By finding a maths teacher who does not.)

  o Explain that this is called disproof by counter-example.

- Write on the board: 'The cube of any number is always positive.'

  o Ask students to suggest values which verify the statement.

  o Prompt students to think of other vales for which it is not true (i.e. negative values).

  o Explain that one case where the statement is not true is enough to disprove a statement.

- Work through Example 11 using ActiveTeach.

Common mistakes and misconceptions

Students often think 'number' means positive whole number – remind them that it represents decimals, negatives, fractions and so on.

Plenary

Split the group into pairs and write the following on the board:
'a is odd and b is even. Therefore $a^2 + b^2$ is odd.'

Ask students in their pairs to decide if the statement is true or false. *If it is true prove it. If it is false give a counter-example.* (True: odd^2 = odd, even^2 = even, odd + even = odd.)

18.1

Mid-point of a line segment

Modular and linear specification reference

N6.3

Keywords

line segment, mid-point, coordinates

Resources

Guided Practice Worksheet

ActiveTeach Resources

Animation

Topic Tutor

Links

Follow up

Middle Practice Book 18.1

Objectives

- Find the mid-point of a line segment **D, C**

Prior knowledge

Students should be able to draw a coordinate grid and plot points in all four quadrants. They should also be able to add and divide negative numbers.

Starter

Display these number pairs:

  a 3, 4   **b** 7, −2   **c** −1, 2   **d** −4, −3

Ask students to work out the average of each pair.

A number pair has an average of 3. One of the numbers is 1. What is the other number?

Main teaching

- Display Example 1 using ActiveTeach.
  - Show that the mid-point occurs at exactly the halfway point between the *x*- (or *y*-) coordinates of the end-points.
  - Students may worry about having to remember the formula. Stress that it is the average of the *x*-coordinates and the average of the *y*-coordinates that they need to find.
  - Remind students that the coordinates of a point can be fractions.
  - Explain to students that if they are not given a diagram they can draw a quick sketch to check that their answer is correct.

Common mistakes and misconceptions

When working out the mid-point, students often subtract the coordinates. Stress that they are finding the average. Using a sketch to check their answer will ensure any errors are highlighted and enable students to re-check their method.

Plenary

A line segment has end-points at (−2, 4) and (−6, 4). What are the coordinates of the mid-point of the line?

A line segment has a mid-point of (6, 3). Find two possible end-points for the line.

M is the mid-point of the line PQ. M has coordinates (−1, 4). Q is the point (−4, 1). What are the coordinates of P?

Plotting straight-line graphs

Objectives

- Recognise straight-line graphs parallel to the *x*- or *y*-axis **E**
- Work out coordinates of points of intersection when two graphs cross **D**
- Plot graphs of linear functions **D, C**

Prior knowledge

Students should be able to substitute positive and negative numbers into simple expressions.

Starter

Display a coordinate grid from −4 to +4. Ask students to plot these points (using different colours for each group):

Group A: (4, 2), (2, 2), (−2, 2)

Group B: (−3, 4), (−3, 0), (−3, −2)

Group C: (3, −1), (0, −1), (−4, −1)

Main teaching

- Display Example 2 using ActiveTeach, and discuss with reference to the Starter activity.

  o Emphasise that a horizontal line has the equation 'y = a number' as it goes straight through that number on the *y*-axis. Similarly, a vertical line has the equation 'x = a number' as it goes straight through that number on the *x*-axis.

  o *Which coordinates will lie on the line $y = 3$? $x = 2$?*

- Display Example 3. Summarise the procedure as:

  Complete the table of values.
  Plot the points.
  Join the points with a straight line.

  o Extend Example 3. *Where does the line $y = 2x + 2$ cross the line $x = 1$? In this instance, the solution can be found from the table of values.* Explain that the solution can also be found by drawing the line $x = 1$ on the same graph and finding the coordinates of the point of intersection.

Common mistakes and misconceptions

Students often find it difficult to calibrate the coordinate axes correctly. Practice is the key here. Although only two points are needed to draw a graph, encourage students to use a third point as a check.

Plenary

Display the graphs of $y = x$ and $y = 2x$. Ask students to read coordinates in all four quadrants. *Is there a link between the x- and y-coordinates? What do you think the graph of $y = 3x$ will look like? $y = 4x$?*

Modular and linear specification reference

N6.4

Keywords

straight-line graph, parallel, *x*-axis, *y*-axis, coordinate pair, linear function

Resources

Guided Practice Worksheet

18.2 Plotting straight-line graphs

ActiveTeach Resources

Animation

Topic Tutors (×2)

Links

Follow up

Middle Practice Book 18.2

Objectives
- Plot straight-line graphs **D, C**
- Find the gradient of a straight-line graph **C**
- Understand the meaning of m and c in the equation $y = mx + c$ **C**
- Find the equation of a line **B**

Prior knowledge

Students should be able to substitute positive and negative numbers into simple expressions, and rearrange simple formulae.

Starter

Display the graphs of $y = x$ and $y = 2x$. Ask students to add the graphs of $y = 3x$ and $y = 4x$. Comment on the steepness of each line.

Main teaching

- Display Example 4 using ActiveTeach.
  - Emphasise that an upward sloping line has a positive gradient, and a downward sloping line has a negative gradient.
- Display the graphs produced in the Starter activity.
  - *What do the equations of the lines have in common? What effect does the number in front of the x have on the graphs?* Emphasise that this number is the gradient of the graph.
  - Add the graph of $y = 2x + 1$ to the grid. *What is different about the equation of this line? What effect does this have on the line?* Elicit that $y = 2x$ and $y = 2x + 1$ are parallel and that the number not linked to x is the y-intercept. Display Example 5.
- Display Example 6.
  - Stress that the operations must be carried out on both sides of the equation. Remind students of the balancing method.
- Display Example 7.
  - Encourage students to use a sketch to help them visualise the line. (See Common mistakes and misconceptions.)

Common mistakes and misconceptions

When finding the equation of a line, students often forget the negative on the gradient. Using a sketch will help them confirm whether the gradient of the line is negative or positive.

Plenary

Display the graphs of $y = 3x$, $y = 3x + 2$, $y = 2x + 4$, $y = x - 2$, $y = 2 - x$ and $y = 3x - 4$, labelled A – F. Ask students to match the equations to the correct line.

Modular and linear specification reference
N6.5h, N6.6h

Keywords
gradient, slope, equation, y-intercept

Resources
none

Guided Practice Worksheet
18.3 Equations of straight-line graphs

ActiveTeach Resources
Animations (×3)
BBC Active video clip
Grade Studio: Problem solving
Topic Tutor

Links
none

Follow up
Middle Practice Book 18.3

18.4 Conversion graphs

Objectives
- Plot and use conversion graphs **E**

Prior knowledge

Students should be able to use direct proportion to work out the cost of items.

Starter

Display: 5 pens cost £2.80.

How much would 10 pens cost? 15 pens? 1 pen? 8 pens?

Main teaching

- Display Example 8 using ActiveTeach.
  - Remind students of the procedure for drawing a graph:

    Complete the table of values.
    Plot the points.
    Join the points with a straight line.

  - Explain that only two points are actually needed to draw a conversion graph but three points provide a good check.
  - Highlight the Common mistakes and misconceptions during the demonstration.
  - *How would you use your answer to part **c** to convert £36 to dollars?*

Common mistakes and misconceptions

Students often find it difficult to read accurately from one value on a conversion graph to find another value. Encourage them to draw horizontal and vertical lines for reading values off the graph.

Plenary

Display the conversion graph for degrees Celsius to degrees Fahrenheit.
- *Why does the graph not go through the origin?*
- *Body temperature is about 37°C. What is this in degrees Fahrenheit?*

Modular and linear specification reference
N6.11

Keywords
conversion graph

Resources
pre-prepared coordinate grid (x-axis from −100 to 100; y-axis from −100 to 240) for Question 2 (for student use if necessary)

pre-prepared conversion graph for degrees Celsius to degrees Fahrenheit (for Plenary)

Guided Practice Worksheet
none

ActiveTeach Resources
Topic Tutor

Links
none

Follow up
Middle Practice Book 18.4

18.5 Real-life graphs

Objectives
- Draw, read and interpret distance–time graphs **E, D, C**
- Sketch and interpret real-life graphs **C**

Prior knowledge

Students should be able to work out what each subdivision on a scale represents.

Starter

Ask students to identify as many units of speed as they can.

Main teaching

- Display Example 9 using ActiveTeach.
  - Emphasise that the *y*-axis is always used for distance and the *x*-axis is always used for time.
  - Discuss what the different sections of the graph represent.
    When was the train travelling fastest? Why?
    Elicit that the steeper line represents a faster speed.
  - Introduce students to the formula triangle for speed. Explain that to calculate an average speed, the total distance of the journey and the total time taken must be used.
- Display Example 10.
  - Link to the Starter activity and discuss the units of speed.
    Why is the speed not given as km/m?
- Display Example 11.
  - Show more examples of vases. Ask students to describe what would be happening to the depth of water as each vase is filled. Then show how the description can be translated to a graph.

Common mistakes and misconceptions

Students often find drawing and labelling axes difficult. Explain that for distance–time graphs they should work out the total distance travelled and the total time taken before deciding on the scale of the axes.

Plenary

Ask students to sketch a graph to represent the depth of water in a bath when the taps are turned on and the plug is in, then the taps turned off, and finally the plug is pulled out.

Modular and linear specification reference
N6.11, N6.12

Keywords
distance–time graph, time, distance, speed, average speed, rate of change

Resources
none

Guided Practice Worksheet
none

ActiveTeach Resources
Animations (×2)
BBC Active video clips (×2)
Topic Tutors (×3)

Links
none

Follow up
Middle Practice Book 18.5

18.6

Using graphs to solve simultaneous equations

Modular and linear specification reference

N5.4h

Objectives

- Use a graphical method to solve simultaneous equations **B**

Keywords

simultaneous equations, intersect

Resources

Guided Practice Worksheet

18.6 Using graphs to solve simultaneous equations

ActiveTeach Resources

Animations (×2)

Links

Follow up

Middle Practice Book 18.6

Prior knowledge

Students should be able to plot graphs of functions in which y is given implicitly.

Starter

Ask students to draw the graph of $y = 2x + 2$.

The line $x = -3$ crosses the line $y = 2x + 2$ at the point A. What are the coordinates of point A?

Main teaching

- Display Example 12 using ActiveTeach.
  - Explain that graphical solutions to simultaneous equations are only approximate as they depend on the accuracy of the drawing.
  - Stress that students should always check their solution by substituting into one of the equations.
  - Emphasise that exam questions should be read carefully – does the question ask for a graphical solution or an algebraic solution?

Common mistakes and misconceptions

Students often forget to ensure that the individual lines are drawn accurately. Remind pupils that while only two points are required to draw a straight-line graph, using a third point provides a good check.

Plenary

Use a graphical method to show why the following simultaneous equations cannot be solved:

$$2x + 2y = 5 \quad \text{and} \quad 3y = 4 - 3x$$

Modular and linear specification reference

N5.7h

Objectives

- Solve inequalities graphically **B**

Prior knowledge

Students should be able to draw straight-line graphs.

Starter

Display a coordinate grid from −10 to +10. Ask students to draw these graphs:

    a $x = 4$ **b** $y = -2$ **c** $y = x$ **d** $y = 3 - x$ **e** $x + y = 4$

Main teaching

- Display Example 13 using ActiveTeach.
  - Summarise the procedure as:

      Convert each inequality to an equation.
      Draw the line.
      Test a point.
      Shade the region.

  - Emphasise that for ≤ and ≥ the boundary is a solid line; and for < and > the boundary line is dotted. (See Common mistakes and misconceptions.)
  - Stress the importance of using a test point to check whether the region above or below the line should be shaded.

- Display Example 14.
  - Stress that the procedure to be followed is exactly the same. The test point is chosen when all three lines have been drawn.
  - Explain that the drawn lines will always enclose the region that students are asked to find.

Common mistakes and misconceptions

Students often mix up whether the lines should be dotted or solid. Emphasise that it is the inequalities containing 'equal to' that are solid.

Students often shade the incorrect area. Using a test point as a check should eliminate this error.

Plenary

Ask students to draw a graph to show the region that satisfies some simple inequalities:

    $y > 2$ $x \leq 4$ $y \geq -4$ $x > y$

Alternatively, display some graphs showing shaded regions and ask students to use inequalities to describe the region.

Keywords

inequality, region,
less than or equal to (≤)
greater than or equal to (≥),
less than (<), greater than (>)

Resources

Guided Practice Worksheet

ActiveTeach Resources

Animation
BBC Active video clip
Grade Studio: Knowledge check
Topic Tutor

Links

Follow up

Middle Practice Book 18.7

Objectives

- Factorise a quadratic expression that is the difference of two squares **B**

Modular and linear specification reference

N 5.2h

Keywords

quadratic expression, difference of two squares, factorise

Resources

Guided Practice Worksheet

ActiveTeach Resources

Topic Tutor

Links

Follow up

Middle Practice Book 19.1

Prior knowledge

Students should be able to expand algebraic expression involving two sets of brackets, and be able to recognise the square numbers.

Starter

Explain that a quadratic expression is one whose highest power of x is x^2. Write the following expressions on the board and ask students to identify the quadratics:

    a $2x + 4$ **b** $4x^2 - 6x$ **c** $9x^2 - 6x^3$ **d** $3x^2 - 7x - 4$

Main teaching

- Ask students to expand these expressions, describing any patterns they notice:

    a $(x + 3)(x - 3)$ **b** $(n + 5)(n - 5)$ **c** $(r - 2)(r + 2)$

    o *Without multiplying out the brackets can you expand the expression:* $(x + 7)(x - 7)$? [$(x^2 - 49)$]

    o Explain that an expression of the form $x^2 - b^2$ is called the 'difference of two squares' because it is the difference (subtraction) between two square numbers.

- Write on the board: $x^2 - 100$.

    o *How might this be factorised?* [$(x - 10)(x + 10)$]

- Display Example 1 using ActiveTeach.

Common mistakes and misconceptions

Mistakes may be made when expanding brackets. For example, students may expand $(x - 3)^2$ incorrectly to give $x^2 - 9$. Remind students to write out $(x - 3)(x - 3)$ and then expand the brackets.

Plenary

Write the following algebraic expressions on the board and ask students to suggest how they might be factorised:

    a $-100 + x^2$ $(-10 + x)(10 + x)$ or $(x - 10)(x + 10)$
    b $49 - x^2$ $(7 - x)(7 + x)$
    c $2x^2 - 200$ $2(x - 10)(x + 10)$
    d $x - 2$ $(x + \sqrt{2})(x - \sqrt{2})$

Objectives
- Factorise a quadratic of the form $x^2 + bx + c$ **B**

Prior knowledge
Students should be able to expand algebraic expression involving two sets of brackets, and also be able to factorise the difference of two squares.

Starter
Ask students to write on their boards the solution to the following:
- *My product is 20, my sum is 12. Who am I?* (2 and 10)
- *My product is –15, my sum is 2. Who am I?* (5 and –3)
- *I think of two numbers. The difference between the smaller and the larger is 4 and their product is 77. What are they?* (11 and 7)

Extend by asking students to write their own problems to try on the class.

Main teaching
- Ask students to expand $(x + 2)(x + 5)$.
  - *What is the coefficient of x in the solution?* (7)
  - *How is this related to the numbers 2 and 5?* (It is the sum.)
  - *What is the numeric value in the solution?* (10)
  - *How is this related to the numbers 2 and 5?* (It is the product.)
- Display: $x^2 + 6x + 8 = (x + \Box)(x + \Box)$
  - Discuss with students which numbers multiply to give 8, listing all the possibilities. (1 and 8; –1 and –8; 2 and 4; –2 and –4)
  - Show students how to work out which numbers are needed by looking at the coefficient of x and deciding which pair sum to this.
  - Explain that if the coefficient of x is positive you will need to have two positive numbers.
- Display Example 2 using ActiveTeach and work through.
- Display Example 3. Establish that if the coefficient of x is negative and the numerical value positive then the two values must be negative.
- Display Example 4. Discuss the fact that if the numerical value is negative, then one of the numbers is positive and the other is negative.

Common mistakes and misconceptions
A common error is looking for numbers whose sum is c and product is b.

Plenary
Display Exercise 19D Q3ai: $x^2 + 6x + 9$. Ask students to factorise this. $[(x + 3)^2]$ Explain that this is a perfect square. Discuss how you can spot a perfect square. (The coefficient of x is double the root of the numeric value.)

Modular and linear specification reference
N 5.2h

Keywords
product, quadratic, factorise, coefficient, sum

Resources
mini-whiteboards

Guided Practice Worksheet
19.2 Factorising quadratics of the form $x^2 + bx + c$

ActiveTeach Resources
Animations (×2)
Topic Tutor

Links
none

Follow up
Middle Practice Book 19.2

Modular and linear specification reference

N5.2h, N5.5h

Keywords

solve, square root, root

Resources

Guided Practice Worksheet

19.3 Solving quadratic equations

ActiveTeach Resources

Topic Tutor

Links

Follow up

Middle Practice Book 19.3

Objectives

- Solve quadratic equations by rearranging **B**
- Solve quadratic equations by factorising **B**

Prior knowledge

Students should know the order of operations, be able to solve linear equations, and factorise a quadratic equation of the form $x^2 + bx + c$.

Starter

Ask students to work out the following (remind them of BIDMAS if needed):

a $3^2 \times 4 + 9 \ (= 45)$ **b** $3(4 + 2^3) - 17 \ (= 19)$ **c** $(5 - 9)^2(8 + 2 \times 4) \ (= 256)$

Main teaching

- Remind students how to solve linear equations.

- Display: $3x^2 + 4 = 100$. Ask students to suggest solutions using trial and improvement techniques. ($x = \pm 5.7$ to 1 d.p.)

  o Discuss how time consuming this was. Explain that there are several ways to solve a quadratic equation. Show students how to solve the equation by rearranging it to make x the subject.

- Display Example 5 using ActiveTeach. Remind students that square roots can be positive and negative, to give two solutions.

- Display Example 6 and work through.

- Display: $2g^2 + 200 = 4g^2$

  o Discuss with students how to solve this, collecting the unknown terms together, and solve to give $g = \pm 10$.

- Display Example 7 and work through.

- Ask students to think of a pair of numbers whose product is zero. Write the pairs on the board and see what they have in common (one is zero). Explain that if the product of two values is zero, one must be zero and that we can use this fact to help us solve a quadratic equation.

- Display Examples 8, 9 and 10 and work through.

Common mistakes and misconceptions

Students often forget there are two solutions to a quadratic equation.

Plenary

Remind students that quadratics always have two solutions.

Display: $x^2 + 4x + 4$ and ask students to find the solution. ($x = -2$)

Modular and linear specification reference

N5.2h, N5.5h

Keywords

Resources

Guided Practice Worksheet

ActiveTeach Resources

Grade Studio: Knowledge check

Grade Studio: Problem solving

Links

Follow up

Middle Practice Book 19.4

Objectives

- Write quadratic equations for problems and then solve them **B**

Prior knowledge

Students should be able to factorise the difference of two squares, and factorise and solve quadratic equations of the form $x^2 + bx + c$.

Starter

Display the following quadratic expressions and ask students to factorise:

- **a** $3x^2 + 6x$ $3x(x + 2)$ or $x(3x + 6)$
- **b** $x^2 + 5x + 6$ $(x + 2)(x + 3)$
- **c** $x^2 - 4x - 5$ $(x - 5)(x + 1)$
- **d** $x^2 - 8x + 12$ $(x - 6)(x - 2)$

Main teaching

- Explain that sometimes it is necessary to form a quadratic equation in order to solve a problem.

- Display the following:
  At a festival an area of field is being fenced off for VIPs. The area enclosed must be 1200 m² and the length must be 10 m longer than the width. What are the dimensions of the area?

  o Ask a student to come to the board and sketch the rectangle.

  o Label the width x.

  o *What is the length of the field?* $(x + 10)$

  o *What is the area of the field?* (1200 m²)

  o *What can you say about the product of the length and width?* (It must be 1200.)

  o Write on the board: $x(x + 10) = 1200$

  o Discuss with students how to solve the quadratic (making one side equal to zero) to give $x = 30$ or $x = -40$.

  o Discuss how only one solution makes sense in this context. (30)

  o *What are the dimensions of the area?* (30 m by 40 m)

- Display Example 11 using Active Teach.

Common mistakes and misconceptions

Students often struggle to identify which is the unknown value or feel there are two and so use two variables.

Plenary

Ask students to design a problem like those in Q3 and Q4 of Exercise 19K.

This short chapter consists of two revision exercises that test students' recall and understanding of number topics covered earlier in the book.

Each exercise contains questions that relate to the objectives below.

The exercises can be done consecutively or separately at different stages.

There are five BBC Active video clips and a Grade Studio: Knowledge check for this chapter.

Objectives

- Understand equivalent fractions
- Simplify a fraction by cancelling all common factors
- Recognise that each terminating decimal is a fraction
- Convert simple fractions to percentages and vice versa
- Use percentages to compare proportions

Common mistakes and misconceptions

When finding equivalent fractions and simplifying fractions, students often forget to multiply or divide both the numerator and denominator. Stress: 'whatever you do to the numerator you must do to the denominator'.

Objectives

- Understand 'reciprocal' as multiplicative inverse

Common mistakes and misconceptions

Students are often unsure what 'reciprocal' means. Constant use of the word will help students form a memory of its meaning.

Objectives

- Use ratio notation

Common mistakes and misconceptions

Students often write the ratio in the incorrect order. Stress that the notation must follow the order given in the question.

Objectives

- Use brackets and the hierarchy of operations
- Add, subtract, multiply and divide integers
- Use calculators effectively and efficiently; know how to use function keys for squares
- Use inverse operations

Common mistakes and misconceptions

Students often forget to use BIDMAS when using calculators to perform calculations. Highlight the importance of having a scientific calculator.

Objectives

- Round to the nearest integer, to one significant figure and to one, two or three decimal places
- Give solutions in the context of the problem to an appropriate degree of accuracy

Common mistakes and misconceptions

Students often treat the digits each side of the decimal point as separate whole numbers, so they round 0.95 to one decimal place to 0.1. Encourage students to consider place value and whether the answer makes sense.

Students often forget to give an answer in the context of the problem. Ensure they check whether their answer makes sense.

21.1 / Angle facts

Modular and linear specification reference
G1.1

Keywords
vertically opposite

Resources
calculators (optional)

Guided Practice Worksheet
none

ActiveTeach Resources
BBC Active video clips: Geometry skills (×1), 21.1 (×1)
Topic Tutor

Links
none

Follow up
Middle Practice Book 21.1

Objectives
- Calculate angles around a point **E**
- Recognise vertically opposite angles **E**

Prior knowledge

Students should be able to solve simple one- and two-step equations. They should also know that angles on a straight line sum to 180°.

Starter

Display a 'spider' diagram with 360 in the centre, and calculations to complete to make 360 round the outside, such as 120 + ☐. Students complete the calculations, then suggest some of their own.

Main teaching

- Face the students and make a full turn. *How many degrees have I turned through?* (360°) Turn again, half-way. *How far have I turned? How much more do I need to turn to get back to the start?* Repeat, turning roughly one third of the way. Illustrate this on the board with an angle of 120° at a point. *What is the remaining angle?*

- Display Example 1 using ActiveTeach and work through, reinforcing the different angle facts.

- Ask students to draw two straight lines that cross and measure the four angles. *What do you notice?* Confirm that vertically opposite angles are equal. Demonstrate how to show this by giving equal angles the same letter, and by using same number of arcs for equal angles.

- Display the three angle facts with diagrams: straight line, at a point, and vertically opposite. Solve a few examples as a class, encouraging students to state which facts they have used each time.

Common mistakes and misconceptions

Students often try to measure rather than calculate angles – keep protractors locked away.

Plenary

Divide the class into teams. Each team member makes an angle problem for the other team. If the team answer correctly they get a point. If the problem cannot be solved (e.g. not enough angle values given) the setter's team loses a point.

21.2 / Angles in parallel lines

Objectives
- Recognise corresponding and alternate angles **D**
- Calculate angles in diagrams with parallel lines **D**

Prior knowledge

Students should recognise parallel lines and vertically opposite angles, and know the sum of angles on a straight line and at a point.

Starter

Display sets of parallel lines in different orientations and demonstrate use of arrowheads to show parallel lines. Sketch a plan view of a table (i.e. a rectangle). *Which pairs of lines on this sketch are parallel? How can we show this?* (Single and double arrows.)

Main teaching

- Ask students to draw a pair of parallel lines on squared or lined paper, using a ruler. Then draw a diagonal line crossing them both. Trace the lines and angles where the diagonal crosses the top line. Then move this tracing down to where the diagonal crosses the bottom line. *What do you notice? Does this work for everyone's diagram?*
- Display two parallel lines and a diagonal. Colour one angle. *Which angle is the same size as this one?* Students use their own diagrams and tracings to decide. Colour it the same colour. Students do the same on their own diagrams.
- Colour the angle vertically opposite to this one the same colour. Hence show that alternate angles are equal.
- Display Example 2 using ActiveTeach, emphasising the need to state the fact used to find each angle.
- Work through Example 3, where students have to use more than one angle fact to solve the problem.

Common mistakes and misconceptions

Students often confuse alternate and corresponding angles. Students can make their own labelled diagrams and refer to them for each problem.

Plenary

Complete Exercise 21C Q4 as a class, if students have not already attempted it. Point out how one angle of each colour makes a straight line, that is, adds up to 180°. *What does this tell us about the angles in the triangles?* (They are also one of each colour, so they sum to 180°.)

Modular and linear specification reference

G1.2

Keywords

parallel, corresponding, alternate

Resources

squared paper, tracing paper, calculators (optional)

Guided Practice Worksheet

21.2 Angles in parallel lines

ActiveTeach Resources

Topic Tutors (×2)

Links

Follow up

Middle Practice Book 21.2

Modular and linear specification reference

G3.1, G3.6

Objectives

- Use three-figure bearing notation **E**
- Measure the bearing from one place to another **E**, **D**
- Plot a bearing **E**, **D**
- Calculate bearings for return journeys **C**
- Draw and interpret scale diagrams to represent journeys **C**

Keywords
bearing

Resources
protractors, maps with north direction shown

Prior knowledge

Students should be able to draw angles and draw lines to scale.

Starter

Give students maps. *Which way is north on the map?* Explain that in maths problems, as in maps, we always draw north straight up the page.

Guided Practice Worksheet
21.3 Bearings

Main teaching

- Display two crosses, to represent a pond and a lake. *A duck flies from the pond to the lake. In which direction does it fly?* Draw the flight path and a north line at the pond. Explain that the bearing for the journey is an angle measured clockwise from north, and the three-figure convention.

- *Who uses bearings to plan journeys?* (Pilots, sailors, orienteers.) Display two more crosses, to represent two towns. A pilot needs to fly from A to B. *On what bearing should she fly?* Emphasise that you draw the north line where the journey starts and measure the angle there.

- Display Example 4 using ActiveTeach.

- Explain bearings greater than 090°. *A ship sails 4 km south then 6 km west. How can we show this on a diagram, using a scale of 1 cm : 2 km?*

- Display a few north lines. *What do you notice?* (All are parallel.) *What angle facts do you know about parallel lines?*

- Work through Example 5.

ActiveTeach Resources
Animation
Grade Studio: Knowledge check
Grade Studio: Problem solving
Topic Tutor

Links
http://www.ordnancesurvey.co.uk/oswebsite/education/pdf/mapreadingmadeeasypeasy2.pdf
A leaflet on map reading, with a map to display (on p7) on a whiteboard, plus games involving compass and coordinates in the map zone.

Common mistakes and misconceptions

Students confuse which angles they should be finding. Emphasise the importance of wording in questions: 'Write down' means they can find the answer on the diagram without any working, 'Work out' means they have to do some calculation.

Follow up
Middle Practice Book 21.3

Plenary

Choose two villages on a map. Work out the bearing to fly from one to the other, and the return bearing. Use the map scale to work out the distance.

22.1 Time and timetables

Objectives
- Solve problems involving times, dates and timetables **E**

Prior knowledge

Students should be able to find fractions of amounts, and write fractions in lowest terms. They should also be able to subtract confidently across tens, hundreds and thousands boundaries.

Starter

Practise converting between hours and minutes. *Write 315 minutes in hours and minutes. What fraction of one hour is 24 minutes?*

Main teaching

- Remind students about using the 12- and 24-hour clock.
  - Explain that you can convert between 12-hour and 24-hour clock times by adding or subtracting 12 to the hours figure.
  - Discuss minute equivalents of common fractions of one hour.
  - Practise finding differences between times.
    How many minutes are there between quarter to 4 and half past 5? Between 12 36 and 13 02?
  - Practise adding on times. *What time is half an hour later than 12:35 am? What time is 45 minutes earlier than 17 12?*
  - Explain that when calculating times that cross an hour boundary it is often easier to complete the calculation in two parts.
- Display Example 1 using ActiveTeach.
  - Explain that timetables usually use the 24-hour clock.
  - Explain that columns 6 and 7 of the timetable give bus times as minutes past the hour, and that the arrows in the first two columns indicate that the bus does not stop at these places.
  - Use real local bus or train timetables to calculate journey times. Ask students real-life questions using timetables. *What time bus should I get from Woodbridge if I want to be in Ipswich for work at 9 am?*

Common mistakes and misconceptions

Students often write 1 hour 25 minutes as 1.25 hours, or vice versa. Remind students to be careful when dealing with fractions and decimals of hours.

Plenary

What's the time? How long until the end of the lesson? Until home time?

Modular and linear specification reference
N1.3

Keywords
12-hour clock, am, pm, midnight, midday, 24-hour clock, hours, minutes, seconds

Resources
local bus or train timetables (optional)

Guided Practice Worksheet
22.1 Time and timetables

ActiveTeach resources
none

Links
http://www.bgfl.org/bgfl/custom/resources_ftp/client_ftp/ks2/maths/timetables/index.htm
An activity for finding the difference between two times.

Follow up
Middle Practice Book 22.1

<table>
<tr><td>

Objectives
- Know and use approximate metric equivalents of pounds, feet, miles, pints and gallons **E**

Prior knowledge

Students should be able to use a calculator to multiply and divide, and should be able to round their answers to a given number of decimal places.

Starter

Rehearse standard metric units of length, capacity and mass. Display a quantity in metres (e.g. 25 m). Choose alternative units (e.g. km) and ask students to convert the quantity. Repeat with different units. Students could write their responses on mini-whiteboards and display them.

Main teaching

- Explain that there are two different systems of units for measuring height, mass and capacity.
  - Explain that you cannot make exact conversions between metric and imperial units. Establish the most common approximations for metric and imperial conversions.
  - Link unit conversions to proportionality. Explain that the $y = kx$ formula can be used to find formulae for unit conversion. Challenge students to find values of k for the conversions given in the explanatory text: miles to kilometres, inches to centimetres, kilograms to pounds, and gallons to litres.
- Display Example 2 using ActiveTeach.
  - *How can we convert from gallons into litres?* (× 4.5)
  - Display '5 miles ≈ 8 km' and '1 mile ≈ ☐ km' on the board. Make the link to equivalent ratios.
  - *How can we convert from miles to km in a single calculation?* (× 1.6)

Common mistakes and misconceptions

Students are often confused about whether they should multiply or divide. Encourage students to use the relative size of units to decide the correct operation (i.e. inches are bigger than cm so you multiply to convert to cm).

Plenary

Display a jar of something like honey or coffee with a mass of 227 g or 454 g. *Why do you think these quantities are chosen?* Ask students to convert these quantities into pounds. *What do you notice?*

</td><td>

Modular and linear specification reference

G3.4

Keywords

imperial units, convert

Resources

mini-whiteboards (for Starter, optional)

jar or packet (or picture) showing a mass of 227 g or 454 g (for Plenary)

Guided Practice Worksheet

22.2 Converting between metric and imperial units

ActiveTeach Resources

BBC Active video clips (×2)

Grade Studio: Problem solving

Links

www.google.com

The Google search engine will convert units directly. For example, enter '12 miles in km' in the search field.

Follow up

Middle Practice Book 22.2

</td></tr>
</table>

Objectives
- Use and interpret maps and scale drawings **E**

Prior knowledge

Students should be able to divide a quantity in a given ratio, and to solve problems involving ratios. They should also be able to confidently convert between different metric units of length.

Starter

Play a 'Pelmanism' game involving equivalent ratios. Draw a grid of ratios on the board containing pairs of equivalent ratios. Cover each ratio with a sticky note. Divide the class into two teams. In turn, teams ask for a pair of sticky notes to be removed. If equivalent ratios are uncovered, they win those two notes. The winning team has the most notes at the end.

Main teaching

- Display a map on the board. If possible provide students or pairs of students with their own copies.
  - *Can you find a scale on the map? What does this tell you?*
  - If the scale is given as a ratio, ask students to write the scale as '1 cm = ☐ km.'
  - If a physical scale is provided, ask students to measure the scale. Challenge students to convert this measurement into a map ratio.
  - Choose points on the map and ask students to measure the distance between them.
  - Ask students to use the map scale to determine the real-life distance between these points.
- Work through Example 3 using ActiveTeach
  - Challenge more able students to work out the area on the map that would represent 1 km² in real life.

Common mistakes and misconceptions

Students should take care when converting between km and cm, and when measurements are given in a variety of units.

Plenary

Discuss a suitable scale to use for a scale drawing of a floor plan of a classroom, or for a map of a town centre. *What are the benefits of using a small scale?* (Can show more detail.) *A large scale?* (Can show greater area.)

Modular and linear specification reference
G3.1

Keywords
scale drawing, map, scale, ratio

Resources
sticky notes (for Starter)
maps or map photocopies

Guided Practice Worksheet
none

ActiveTeach Resources
BBC Active video clips (×2)
Topic Tutor

Links
none

Follow up
Middle Practice Book 22.3

Objectives
- Recognise that measurements given to the nearest whole unit may be inaccurate by up to one half unit in either direction **C**

Prior knowledge

Students should be able to add and subtract decimal numbers. They should also be able to round to a given place value or number of decimal places.

Starter

Display a 5-digit number on the board, such as 34 000. *This number has been rounded to the nearest 1000. Give an example of what the original number might have been.* Display a similar number such as 65 000. *This number has been rounded to the nearest 100. Give an example of what the original number might have been.* Repeat with decimal numbers.

Main teaching

- Ask if any students know their heights. Record these heights on the board.
  - *How exact are these measurement? Is Harriet exactly 146 cm tall? Exactly 5 feet 4 inches tall? What is the tallest/shortest possible value for her exact height?*
  - Explain that if a measurement is given to the nearest whole unit it might be inaccurate by up to half a unit in either direction.
  - Explain that the maximum and minimum possible values for the actual measurement are called the upper bound and lower bound.
- Work through Examples 4 and 5 using ActiveTeach.
  - Explain that if you let the mass of the dog in Example 4 be m kg you can write the range of possible values as $33.5 \leq m < 34.5$.
  - Explain that quantities rounded to a number of decimal places or the nearest 10, 100 or 1000 also have upper and lower bounds.

Common mistakes and misconceptions

Students are often confused about the definition of the upper bound, since for example, 146.5 rounds to 147. Explain that 146.5 is the only possible value for the upper bound: for any smaller value you can always find a larger value which still rounds to 146.

Notes on some problem-solving questions

In Exercise 22D Q4, students must use the upper bounds of both values when calculating the maximum area.

Plenary

Discuss upper and lower bounds of discrete quantities. *Rounded to the nearest hundred, why is the upper bound of 3500 people different from the upper bound of 3500 metres?*

Modular and linear specification reference

G3.3

Keywords

lower bound, minimum value, upper bound, maximum value

Resources

Guided Practice Worksheet

ActiveTeach Resources

Animations (×2)
BBC Active video clips (×2)
Grade Studio: Knowledge check
Topic Tutor

Links

http://www.waldomaths.com/ video/Bounds01/Bounds01.jsp
A video explaining upper and lower bounds.

Follow up

Middle Practice Book 22.4

Modular and linear specification reference

G1.1, G1.2

Keywords

interior angle, exterior angle

Resources

Guided Practice Worksheet

23.1 Interior and exterior angles

ActiveTeach Resources

Animations (×2)

Topic Tutor

Links

Follow up

Middle Practice Book 23.1

Objectives

- Solve angle problems in triangles **E**
- Solve angle problems in triangles involving algebra **D**

Prior knowledge

Students should know the names and properties of the different types of triangles: right-angled, scalene, isosceles and equilateral.

Starter

Ask the students some quick-fire questions. Tell them an angle (such as 75°), and they have to tell you the supplement of that angle (105° is the supplement of 75°).

Main teaching

- Students may need revision of the properties of triangles, especially the differences between isosceles and equilateral triangles.
- Students must know, or learn, the following:
  - o angles in a triangle add to 180°
  - o angles on a straight line add to 180°.
  - o It is also useful for the students to know that the exterior angle is equal to the sum of the two opposite interior angles.
- If students have not done so before, a hands-on run through of the explanatory text on interior and exterior angles is advisable.
- Display Example 1 using ActiveTeach.
  - o Work through the example step by step.

Common mistakes and misconceptions

Students may not realise when a triangle is isosceles and so think that the problem cannot be solved. When answering the algebra-based questions, mistakes are often made by trying to do too many steps in one go.

Plenary

Discuss the different methods of working out all the angles in these triangles. Ask students to work out the answers.

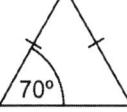 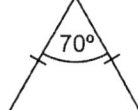 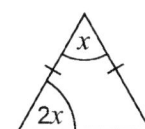 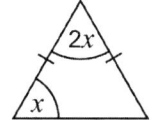

Modular and linear specification reference
G3.9, G3.10

Keywords
construct, arc

Resources
compasses, ruler, protractor

Guided Practice Worksheet
none

ActiveTeach Resources
Animation
Grade Studio: Problem solving
Topic Tutor

Links
none

Follow up
Middle Practice Book 23.2

Objectives

- Draw triangles accurately when given the length of all three sides **E**
- Draw triangles accurately when at least one angle is given **D**

Prior knowledge

Students need to be able to draw and measure straight lines, circles and angles accurately.

Starter

Ask the students to draw the following diagrams; the size of the angle doesn't matter. Once they have drawn the diagrams, get them to measure the angles and write down their measurements.

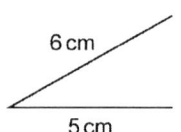 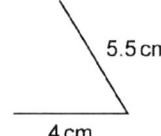 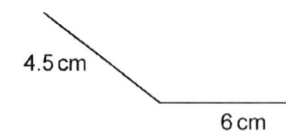

Get students to swap books with a friend. Ask them to measure the lines and angles on their friend's diagrams to check they are accurate. Allow an error of ± 2 mm on the lines and ± 2° on the angles.

Main teaching

- Display Example 2 using ActiveTeach.
  - Ask students to follow the instructions to draw the triangle.
  - Ask them to measure $\angle CAB$ on their diagrams. Answers between 44° to 52° means they have drawn the triangle accurately.
- Display Example 3.
  - Ask students to follow the instructions to draw both triangles.
  - Ask them to measure the third side of triangle **a** on their diagrams; 5.1 cm to 5.7 cm means they have drawn the triangle accurately.
  - Ask them to measure the third angle in triangle **b** on their diagrams. Answers between 66° to 74° means they have drawn the triangle accurately.

Common mistakes and misconceptions

Inaccurate use of the protractor or compasses will cause mistakes.

Students may not complete the triangle by drawing the third side.

Plenary

Ask students to draw this shape, then measure the length of side AB. Their answer should be between 6.2 cm and 6.9 cm.

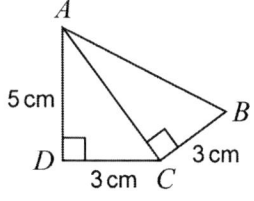

23.3 / Congruent triangles

Objectives
- Recognise and explain how triangles are congruent **C**

Prior knowledge

Students need to be able to identify shapes that are identical.

Starter

Draw on the board the following triangles. Ask students if the shapes are identical. Explain that if they are identical, they are called congruent.

Draw on the board the following triangles. Ask students if the shapes are identical. Discuss the fact that they can still be identical even if they have been reflected or rotated.

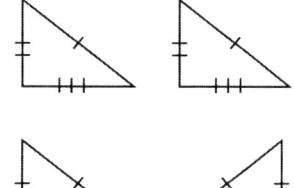

Main teaching

- Using ActiveTeach, display the explanatory text on congruency of triangles.
  - Work through the four conditions for congruency carefully.
- Display Example 4.
  - Work through both parts, making sure that students understand why the two triangles in part **b** aren't congruent.

Common mistakes and misconceptions

Students often make the mistake of thinking that two triangles are congruent when they are not (due to the relative positions of side lengths or angles being in different positions). Encourage them to look very carefully at the relative positions of the angles and side lengths.

Plenary

Ask students which question in Exercise 23D they found the most difficult, and why. Discuss how they decided on the 'odd one out' in Q3.

Modular and linear specification reference
G1.8

Keywords
congruent, included angle

Resources
none

Guided Practice Worksheet
none

ActiveTeach Resources
Grade Studio: Knowledge check

Links
none

Follow up
Middle Practice Book 23.3

Objectives

- Write your own formulae and equations **D**
- Substitute into a formula to solve problems **D, C**
- Set up and solve equations **D, C, B**
- Change the subject of a formula **C, B**

Prior knowledge

Students should be able to solve linear equations with two operations and the variable on both sides, and be able to solve quadratic equations by factorising. They should also be able to confidently substitute values into formulae.

Starter

I think of a number. I add 12, multiply by 2 then subtract 5. I end up with 31. What number did I think of? (6)

Discuss strategies for solving this problem (i.e. trial and improvement, inverse operations, writing and solving equations).

Main teaching

- Display Example 1 using ActiveTeach.
  - *What do the angles in a triangle add up to?* (180°)
  - Explain that you can represent the information given in the question as an equation.
  - Suggest that students write their equation in the simplest way possible (in this case $a + a + 26 = 180$), then simplify and solve.
  - Rehearse factorising and solving quadratic equations.
- Display Example 2.
  - Remind students that a formula doesn't have an answer, but that it is a rule for finding values. Explain that if you want to solve the same type of problem many times writing your own formula can save time.
- Display Example 3.
  - Demonstrate the parallel processes of rearranging the formula and solving the equation $42 = 24 + 2w$.

Common mistakes and misconceptions

Students often fail to consider different terms of an expression when changing the subject of a formula (e.g. $W = \frac{1}{2}x + 3 \Rightarrow 2W = x + 3$). Ask students to solve a similar equation using the desired subject of the formula as the unknown, writing down the operations used.

Plenary

Challenge students to create their own angle problem or word problem which can be solved using algebra. Challenge more able students to create problems with fractional or negative answers.

Modular and linear specification reference

N5.4, N5.6

Keywords

equation, formula, substitute, solve, subject

Resources

Guided Practice Worksheet

24.1 Equations and formulae

ActiveTeach Resources

Topic Tutors (×3)

Links

http://www.mathplayground.com/algebra_puzzle.html
A puzzle which can be solved by forming and solving equations.

Follow up

Middle Practice Book 24.1

Modular and linear specification reference
G2.3, G2.3h

Keywords
proof, prove

Resources
paper, scissors for teacher and students (optional)

Guided Practice Worksheet
none

ActiveTeach Resources
Animation
Grade Studio: Problem solving

Links
none

Follow up
Middle Practice Book 24.2

Objectives
- Prove simple results from geometry **C**

Prior knowledge

Students should know angle facts for triangles, lines, points and parallel lines, and be able to use them to find missing angles.

Starter

Display a diagram containing parallel lines, such as the one shown. Ask students to come to the front and identify a pair of opposite angles, alternate angles or corresponding angles.
What do we know about these pairs of angles?
(They are equal.)

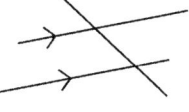

Main teaching

- Cut a large triangle out of paper. Instruct students to cut their own triangle out of paper.
  - o Instruct students to tear their triangle into three corners, and assemble the corners as shown in Example 4. *What do you notice?*
  - o *Does this prove that the angles in a triangle add up to 180°?*
  - o Explain that a proof in maths must work for every possible triangle. *Are you convinced that this is true for every triangle? Could you test it for every possible triangle?*
- Display Example 4 using ActiveTeach.
  - o Carefully explain each step of the proof.
  - o Explain that for each step of the proof you have to write down which angle fact you have used.
  - o Explain that a proof must make sense to someone who is reading it. The whole proof is the answer.

Common mistakes and misconceptions

Students often struggle to lay out their answers neatly and provide reasons for each stage of their working. Suggest that students write the five angle facts on a piece of paper or card so that they can refer to them easily. Remind students that every time they use an angle fact they must write it down using the correct name.

Plenary

Ask a student or pair of students to come to the front of the class and recreate the proof that the angles in a triangle total 180° without referring to their books. Encourage other students to fill in any missing stages.

Objectives

- Use trial and improvement to find solutions to equations **C**

Prior knowledge

Students should be able to use a calculator confidently to complete complex calculations. They should also be able to order decimals and find the value half way between two numbers.

Starter

Play 'Too big or too small'. Secretly write down a number between 1 and 100. Ask students to guess the number. Tell them whether they are too big or too small. Students should refine their guesses and try to find the number in the fewest number of trials possible. Record trials on the board. Discuss different students' strategies.

Main teaching

- Write this equation on the board: $x^3 - x = 50$

  - *Does anybody know how to solve this equation?* Remind students that a solution to an equation is any value for x which makes the equation true.

  - Explain that if you can't find an exact solution to an equation using algebra you can use trial and improvement.

  - Display a table of values or spreadsheet with columns for x, $x^3 - x$ and Comment. Invite students to suggest solutions and determine whether they are too small or too big. Suggest strategies for saving time (i.e. choosing a closer trial if the solution is very close).

- Display Example 5 using ActiveTeach.

  - Explain that if you want to find the solution correct to one decimal place then knowing it is between 3.3 and 3.4 is not enough.

  - Demonstrate the use of the mid-point (3.35) to determine the correct answer.

Common mistakes

Students may make a choice between, for example, $x = 3.3$ and $x = 3.4$ based on the value of the function and the desired output. Reinforce the need to check the mid-point to determine which value is correct.

Plenary

Display a variety of equations including equations involving fractions, quadratics and cubics. Discuss which equations students are able to solve using algebra and which could be solved using trial and improvement.

Modular and linear specification reference

N5.8

Keywords

trial and improvement

Resources

spreadsheet application on whiteboard (optional)

Guided Practice Worksheet

24.3 Trial and improvement

ActiveTeach Resources

Grade Studio: Knowledge check

Topic Tutor

Links

http://www.tes.co.uk/article.aspx?storycode=6017925

An activity with accompanying spreadsheet using a numerical method to solve a problem. (Note that tes.co.uk requires free registration.)

Follow up

Middle Practice Book 24.3

Modular and linear specification reference

G1.2, N5.4

Keywords

quadrilateral, diagonal

Resources

Guided Practice Worksheet

25.1 Quadrilaterals and algebra

ActiveTeach Resources

Animations (×2)

Topic Tutor

Links

Follow up

Middle Practice Book 25.1

Objectives

- Calculate interior angles of quadrilaterals **E**
- Solve angle problems in quadrilaterals involving algebra **D**

Prior knowledge

Students should know that angles on a straight line add to 180° and that angles in a triangle add to 180°.

Starter

Ask students quick-fire questions involving complements to 180 and 360.

- Say a number such as 150, 40, 125, etc.; they have to tell you the number they must add to your number to make 180.
- Say a number such as 210, 60, 255, etc.; they have to tell you the number they must add to your number to make 360.

Main teaching

- Show students that the sum of the interior angles of a quadrilateral (and indeed all polygons) can be found by dividing it into triangles.
- Display Example 1 using ActiveTeach.
  - Work through the example, reminding students that they should write down the reasons for their calculations. This may be an appropriate time to ask students for any other angle facts that they can recall.
- Display Example 2.
  - Work through the example, step by step.

Common mistakes and misconceptions

Students often try to work things out in their heads without writing down the calculations they are trying to do.

Students will lose method marks if they do not show full workings for algebra questions if they make mistakes.

Plenary

Ask students to find the value of x in this diagram. Tell them they must show all of their workings. Once they have finished, ask them to swap books with a friend. The friend must mark their work and point out any steps of the workings that have been missed.

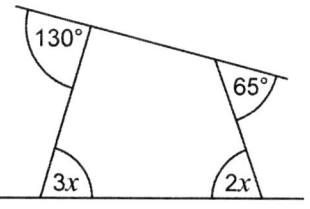

25.2 / Special quadrilaterals

<table>
<tr><td>

Objectives
- Make quadrilaterals from two triangles **E**
- Use parallel lines and other angle properties in quadrilaterals **D**

</td></tr>
</table>

Prior knowledge

Students should know the names of the different types of triangles, and should know angle facts such as: $90°$ in a right angle; angles on a straight line add to $180°$; angles in a triangle add to $180°$; there are $360°$ in a full turn.

Starter

Use the Skills check to brush up on the prior knowledge needed to work through this section.

Main teaching

- Display and work through the explanatory text on the names of special quadrilaterals. Give special emphasis to the properties of parallel sides and equal angles of the quadrilaterals.
  - Remind students of the angle properties when dealing with parallel lines, that is, alternate, corresponding and allied angles.
- Display Examples 3 and 4 using ActiveTeach.
  - Work through the examples, emphasising that students should write down the properties they have used, as shown in the brackets in Example 4.

Common mistakes and misconceptions

Students may give correct answers, but do not explain the properties used.

Plenary

Discuss Q4 and the various methods used by students to get the answers required.

Modular and linear specification reference

G1.2, G1.4

Keywords

square, rectangle, rhombus, bisect, parallelogram, trapezium, kite, adjacent

Resources

Guided Practice Worksheet

ActiveTeach Resources

Animation

BBC Active video clip

Links

Follow up

Middle Practice Book 25.2

25.3 | Polygons

Objectives

- Use the exterior angles of polygons to solve problems **D**, **C**
- Solve more complex angle problems involving exterior and interior angles of a polygon **D**, **C**

Prior knowledge

Students need to be able to solve simple equations.

Starter

Ask students to sketch an equilateral triangle and a square with sides extended as shown here.

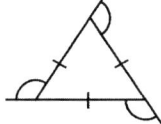 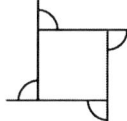

For each shape, ask students to write down the sizes of the angles shown. Discuss with them whether or not it is a coincidence that the exterior angles of each shape are the same, and whether this would work with other polygons.

Main teaching

- Students must learn that the sum of the exterior angles of any polygon is 360°.
- Students also need to learn or deduce that the number of sides of a regular polygon is 360 ÷ the exterior angle.
- Display Examples 5 and 6 using ActiveTeach.
  - Work through the examples, then ask the students if they know the name of a 9-sided polygon. (Nonagon)
- Display Examples 7 and 8.
  - Work through the examples, emphasising the different approaches used for a hexagon and a regular hexagon.

Common mistakes and misconceptions

Some students will incorrectly split the polygon into triangles. For example, they will draw [pentagon figure] instead of [pentagon figure]

Plenary

Ask students to work out the exterior angle and interior angle of a regular dodecagon (12 sides). (Answers: 30°, 150°.)

Modular and linear specification reference

G1.3, N5.4

Keywords

polygon, regular, exterior angle, interior angle

Resources

Guided Practice Worksheets

25.3a Polygons 1
25.3b Polygons 2

ActiveTeach Resources

Animation
BBC Active video clip
Grade Studio: Problem solving
Topic Tutor

Links

Follow up

Middle Practice Book 25.3

Modular and linear specification reference

N6.3

Keywords

Resources

Guided Practice Worksheet

ActiveTeach Resources

Grade Studio: Knowledge check

Links

Follow up

Middle Practice Book 25.4

Objectives

- Plot all points of a quadrilateral given geometric information **E**
- Find the mid-point of a line plotted on the coordinate axes **E**

Prior knowledge

Students need to be able to draw and identify special quadrilaterals. They should also be able to plot points in all four quadrants of a coordinate grid.

Starter

Ask students to tell you the number that is half way between pairs of numbers such as:

2 and 4	2 and 5	1 and 7	0 and 3
0 and 1	0 and –1	0 and –2	–1 and –3
–1 and –2	–1 and 1	–1 and 2	–1 and 3

Main teaching

- Students may need to be reminded of the names and properties of some of the special quadrilaterals.
- Display Example 9 using ActiveTeach.
  - Work through the example step by step and supplement by discussing how to work out the mid-points of both AD and AC. [$(-\frac{1}{2}, 3)$ and $(-\frac{1}{2}, \frac{1}{2})$ respectively]

Common mistakes and misconceptions

Students may plot the numbers on the x- and y-axes the wrong way round.

Students may not be able to recognise, or name, some of the less common quadrilaterals such as the kite and trapezium.

When finding the mid-points, students may average only the x- or y-coordinate and not both.

Plenary

Ask students to make a copy of the grid in Q3. Then ask them to plot these points: $(-4, -1)$, $(-4, -4)$, $(-1, -4)$.

- *What are the coordinates for the fourth corner of a square?* $(-1, -1)$
- *What are the coordinates for the fourth corner of an arrowhead?* $(-3, -3)$
- *How many coordinates can you find for the fourth corner of a kite?* (7 on the grid drawn)
- *How many coordinates can you find for the fourth corner of a trapezium?* (16 on the grid drawn)

Objectives

- Find the perimeter and area of rectangles, parallelograms, triangles and trapezia **E, D**

Prior knowledge

Students should know how to convert between centimetres and metres, and how to multiply numbers with up to one decimal place.

Starter

Estimate $2 \times 11.3 + 2 \times 5.9$, *then calculate exactly. How close is the estimate?*
(Estimate = $2 \times (11 + 6) = 34$; exact answer is 34.4)

Estimate $\frac{1}{2} \times 14.8 \times 42.3$, *then calculate exactly.* (Estimate = 300; exact answer is 313.02)

Why are estimated answers useful?

Main teaching

- Display the explanatory text using ActiveTeach to demonstrate the derivation of perimeter and area formulae for rectangle, parallelogram and triangle.

- Display Example 1 and work through, explaining how to calculate the perimeter and area of different shapes.

- Depending on the ability of the class, you may wish to go through one of the parts of Exercise 26A Q2 to show how to calculate lengths of shapes when the area is known.
  For example, in part **b**, if $2.5 \times y = 30$ cm, what is the value of y?
  - Divide both sides by 2.5 to get y on its own: $y = \frac{30}{2.5} = 12$
  - *Don't forget the unit. If area is cm² and the known length is in cm, then y is in cm. So y = 12 cm.*

- Show how to find the area of a trapezium. Explain that students do not need to remember how to derive the formula, or even to remember it, but they must be able to recognise a trapezium shape.

Common mistakes and misconceptions

Encourage students to make a rough estimate of areas and perimeters as a check to avoid arithmetical errors.

Students may make mistakes when converting between units.

Plenary

Draw an 8 × 8 square and a 4 × 16 rectangle. *Do these shapes have the same area?* (Yes; 64 sq. units.) *Do they have the same perimeter?* (No; 32 units, 40 units.)

Challenge students to find other shapes with the same area but different perimeter.

Modular and linear specification reference
G4.1

Keywords
perimeter, area, parallelogram, perpendicular height, trapezium

Resources
none

Guided Practice Worksheet
26.1 Perimeter and area of simple shapes

ActiveTeach Resources
Animations (×3)
BBC Active video clip
Topic Tutors (×2)

Links
http://www.ncetm.org.uk/Resources/15811
http://nrich.maths.org/6398

Follow up
Middle Practice Book 26.1

Objectives

- Find the perimeter and area of compound shapes **E, D**

Prior knowledge

Students should know how to find unknown lengths if given one length and the perimeter or area of a rectangle.

Starter

Revise order of operations. *Work out:*

   a *3 + 5 + 10 + 9 + 2 + 4 + 10* (43)

   b *2 × 7 + 2 × 6* (26)

   c *7 × 6 − 5 × 4* (22)

Main teaching

- Explain that compound shapes are shapes made from joining simple shapes together and that we follow the same rules for working out perimeters and areas as for simple shapes.
- Draw some compound shapes. Ask volunteers to show how the shape could be split into several simple shapes. Make these points:
  - there is often more than one way of splitting up the shape
  - you can use existing measurements to find any missing ones
  - you can think of the perimeter as a route around the shape
  - you need to know whether to add or subtract the individual areas
  - don't forget to include the correct units in your answer.
- Display Examples 3 and 4 using ActiveTeach and work through.

Common mistakes and misconceptions

Students sometimes calculate missing lengths incorrectly, or add areas instead of subtracting.

Plenary

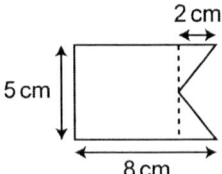

Ann calculates the area as: $8 \times 5 + \frac{1}{2} \times 2 \times 5 = 45$ cm².

Bindi calculates: $8 \times 5 - 2 \times 5 = 30$ cm².

Carl works out: $8 \times 5 + 2 \times 5 \times \frac{1}{2} = 8 \times 7 \times 5 \times \frac{1}{2} = 140$ cm².

What is the correct answer? (35 cm²) *What has each student done wrong? How can you easily tell that Ann and Carl have got their answers wrong?* (By estimating.)

Modular and linear specification reference

G4.1

Keywords

compound shape

Resources

Guided Practice Worksheet

ActiveTeach Resources

Animation

BBC Active video clip

Topic Tutor

Links

http://www.ncetm.org.uk/Resources/16356

Follow up

Middle Practice Book 26.2

Modular and linear specification reference

G4.1, G4.4

Keywords

prism, cross-section, cuboid

Resources

Plasticine or alternative moulding material, centicubes

Guided Practice Worksheet

26.3 Volume and surface area of prisms

ActiveTeach Resources

Grade Studio: Knowledge check

Grade Studio: Problem solving

Topic Tutor

Links

http://nrich.maths.org/4919

Follow up

Middle Practice Book 26.3

Objectives

- Find the volume and surface area of a prism **E, D, C**

Prior knowledge

Students should know how to find areas of triangles, parallelograms and trapezia.

Starter

Suppose you want to wrap a present. Will the wrapping paper be big enough to fit the present? Could you calculate it? What do you need to know? Volume of present? Establish that it is more useful to know the area of the wrapping paper and the surface area of the present. Use a cuboid of Plasticine to show that changing the shape doesn't change the volume, but can change the surface area: some shapes are easier to wrap than others.

Main teaching

- Show how a cuboid is made up of centicubes and illustrate that by counting the layers, we can find the total volume.

- The volume of regular shapes such as cuboids can be found by using a formula such as $l \times w \times h$. But $w \times h$ is also the area of the face.

- Extend this idea to other prisms, reinforcing the meanings of the terms prism and cross-section.

- Display Example 5 using ActiveTeach, showing how to calculate volumes using lengths and cross-sections.

- Use ActiveTeach to display the explanatory text on unfolding a cuboid to make a net. Make the following points:

  o adding the areas of the net rectangles gives the total surface area

  o the faces of a cuboid are in pairs, so there is often a short cut to calculating surface areas

  o the face of a prism may itself be a compound shape.

  o Encourage students to sketch the net of the shape and number or letter all the faces so they don't double count or miss out a face.

- Display and work through Example 6.

Common mistakes and misconceptions

Students may mix up volumes and surface areas, and work out surface area when asked for volume and vice versa.

Plenary

The volume of a triangular prism is 216 cm³. If the side length of the prism is 8 cm, what is the area of the triangular face? (27 cm²) What is the side length of a cube that has the same volume? (6 cm)

Suggest other prisms that have the same volume.

Objectives
- Make a drawing of a 3-D object on isometric paper **E**
- Draw plans and elevations of 3-D objects **D**
- Identify planes of symmetry of 3-D objects **D**

Prior knowledge

Students should know how to draw nets of 3-D objects.

Starter

What is meant by a bird's eye view? Open up a mapping website (see Links) and select your local area. What features can students recognise from the plan view?

Main teaching

- Show the structure of isometric paper and go through Example 1 using ActiveTeach. Emphasise that:
  - the cross-section is drawn first
  - the measurements on the 3-D view can easily be transferred to the isometric drawing by counting the vertices.
- Make up a model using multilink cubes, and demonstrate the plan and elevations by rotating the model.
  - Show how plans and elevations can be used to work out surface areas. Write in ×2 next to any plan or elevation that is repeated. This reduces numbers of drawings needed but ensures that an area is not forgotten when calculating the total surface area.
- Display and work through Examples 2 and 3.
- Revise lines of symmetry in 2-D objects; ask some volunteers to draw a 2-D shape and show one or more lines of symmetry. Ask the rest of the class to say if they are correct and whether any more can be added.
- Move on to explain that some 3-D objects have plane symmetry.
  - As for line symmetry, one half must be an exact mirror image of the other half. There may be more than one plane of symmetry.
- Demonstrate Example 4.

Common mistakes and misconceptions

Students may miss out hidden cubes when converting from a 3-D view to a plan or elevation.

Plenary

Show a model made from 5 multilink cubes. Ask half the class to draw it on isometric paper. The other half of the class should draw the plan view, and front and side elevations.

Modular and linear specification reference
G2.4

Keywords
cross-section, plan, front elevation, side elevation, plane of symmetry

Resources
isometric paper, multilink cubes

Guided Practice Worksheet
none

ActiveTeach Resources
Animation
BBC Active video clip
Grade Studio: Knowledge check
Grade Studio: Problem solving
Topic Tutor

Links
http://www.multimap.com/tour/map/01/ (for Starter)
http://www.mathsnet.net/geometry/solid/guessview.html
http://nrich.maths.org/787 (for Main teaching)

Follow up
Middle Practice Book 27.1

Modular and linear specification reference

G1.7

Objectives

- Draw reflections on a coordinate grid **E, D, C**
- Describe reflections on a coordinate grid **D, C**

Prior knowledge

Students should be able to reflect objects in vertical, horizontal and diagonal mirror lines on a squared grid.

Starter

Display a coordinate grid in four quadrants. Draw on vertical and horizontal lines, such as $x = 3$, $x = -4$, $y = 2$, $y = -5$, plus the lines $y = x$ and $y = -x$. Challenge students to 'name that line' – include the two axes.

Main teaching

- Discuss students' drawings of reflections in the Skills check and establish that the object and image are congruent, and the same perpendicular distance from the mirror line, on opposite sides.

- Display Example 1 using ActiveTeach and work through, drawing the mirror line on a coordinate grid and then the reflection.

- Work through Example 2, which gives a shape on a grid with its reflection. *Where is the mirror line? What is its equation?* Establish that to describe a reflection on a coordinate grid you need to give the equation of the mirror line. Display 'Reflection in the line _____' as a reminder.

- Give students coordinate grids and ask them to draw a shape and a (faint) mirror line, reflect their shape and then rub out the mirror line. They swap grids and find the mirror line and its equation.

Common mistakes and misconceptions

Students find reflection in diagonal lines difficult. Use mirrors, or trace the shape and line and fold along the mirror line to see where the reflection will be (useful in exams, where they will have tracing paper).

Plenary

Display a coordinate grid. Show students a symmetrical plane shape, such as a square, triangle or kite. *What shape do you need to draw so that it reflects to make this shape? Is there more than one way to do it? Describe the transformation you use.* Encourage them to use all lines of symmetry, so, for example, for a square they can draw a rectangle and reflect it in a vertical or horizontal line, or a triangle and reflect it in a diagonal line.

Keywords

reflection, object, image, congruent

Resources

squared paper (for Skills check)

tracing paper (optional), copies of coordinate grids (optional, to save time copying diagrams in the Exercises)

symmetrical plane shape (for Plenary)

Guided Practice Worksheet

28.1 Reflection on a coordinate grid

ActiveTeach Resources

Animation

Topic Tutor

Links

Follow up

Middle Practice Book 28.1

Objectives
- Translate a shape on a grid **D**
- Use column vectors to describe translations **C**

Prior knowledge

Students should be able to use coordinates in all four quadrants, and should understand positive and negative movements in the x- and y-directions.

Starter

Give students a coordinate grid on cm² paper and a cut-out shape. Give an instruction such as 'translate 3 squares up', and students slide the shape into the new position. Extend to more complex translations.

Main teaching

- Carrying on from the Starter, ask students to place their shape and draw round it. Then they translate it, for example, 4 right and 2 down and draw round it in the new position. Colour one of the vertices on the original shape and on the image. *Describe how this point has translated.* Repeat for another vertex. *What do you notice?* Confirm that all points move the same distance in the same direction.

- Display Example 3 using ActiveTeach.

- Discuss students' answers to Exercise 28C Q5 and Q6, establishing that object and image are congruent, and that if the translation A to D is x across, y down, the translation D to A is $-x$ across, y up.

- Display a shape drawn on a coordinate grid. Carry out different translations, emphasising movement in x-direction and in y-direction, both positive at first, then extending to negative. Demonstrate how to write these movements as a column vector.

- Give students practice in translations given by column vectors, by using shapes and grids from the Starter.

- Work through Example 4, using column vectors to describe translations.

Common mistakes and misconceptions

Students forget what the two values in the column vector mean. Emphasise that x comes first in the alphabet, so x is on top.

Plenary

Display several reflections and translations, and some where the image is not congruent to the object. *Which is a reflection? Which is a translation? How can you tell?* (Congruent; reflection is 'flipped'; translation faces same way.)

Modular and linear specification reference

G1.7, G5.1

Keywords

translation, congruent, column vector

Resources

shapes cut out of card, to fit on a cm² grid with all vertices at grid points, plus cm² paper (for Starter)

copies of coordinate grids to save time copying in Exercises; shape drawn on coordinate grid (for Main teaching)

coordinate grids with reflections and translations (some with image not congruent to object; for Plenary)

Guided Practice Worksheet

ActiveTeach Resources

Animation

Grade Studio: Problem solving

Topic Tutor

Links

Follow up

Middle Practice Book 28.2

28.3 / Rotation

Objectives
- Draw the position of a shape after rotation about a centre **D**, **C**
- Describe a rotation fully giving the size and direction of turn and the centre of rotation **D**, **C**

Prior knowledge

Students should know how many degrees there are in full, half and quarter turns, and be able to apply these turns clockwise and anticlockwise.

Starter

Give students an isosceles triangle cut from card. Rotate this about different points on the triangle. *Which point do you need to rotate about so that the triangle and its image make a rhombus?* (Mid-point of short side.) *Parallelogram?* (Mid-point of longer side.)

Main teaching

- Carrying on from the Starter, establish that the position of the image depends on the position of the centre of rotation.
- Demonstrate how to use tracing paper to draw rotations.
- Display Example 5 using ActiveTeach, where the centre of rotation is not on the shape.
- Display a shape and its rotation on a squared grid. *In which direction has this been rotated?* Demonstrate how to use tracing paper and trial and error to find the centre of rotation, as in the explanatory text.
- Ask students to draw a triangle on a coordinate grid, as in Example 6. Tell them to rotate the shape. Establish that they need to know the angle, direction and centre before they can rotate it. Display this for reference. Work through Example 6.

Common mistakes and misconceptions

Students have difficulty working out the angle of rotation. On a coordinate grid, they can trace part of the y-axis. Then when they rotate, when this line is horizontal they have rotated 90°, when vertical 180°, etc.

Plenary

Give students plane shapes, such as trapezia, parallelograms, triangles. *What shapes can you make by rotating around vertices and mid-points?*

Modular and linear specification reference
G1.6, G1.7

Keywords
rotation, centre of rotation

Resources
isosceles triangles cut from card (for Starter)
tracing paper, squared paper, coordinate grids to save time copying, shape and its rotation on a squared grid (for Main teaching)
other plane shapes cut from card (for Plenary)

Guided Practice Worksheet
28.3 Rotation

ActiveTeach Resources
Animation
BBC Active video clips (×3)
Topic Tutor

Links
none

Follow up
Middle Practice Book 28.3

Objectives
- Transform shapes using more than one transformation **C**
- Describe combined transformations of shapes on a grid **C**

Prior knowledge

Students should understand reflection, rotation and translation, and be able to use column vectors.

Starter

Ask students to draw a shape in one quadrant of a coordinate grid and label it A, then carry out the same transformations as in the Skills check. They swap grids with a partner. *What can you say about all the shapes?* (They are congruent.) *How can you tell which shape is a translation/ reflection/rotation of A?* (Shape faces same/opposite/different direction.)

Main teaching

- Explain that combined transformations simply means carrying out one transformation, then another. Ask students to draw a triangle A on a coordinate grid and translate it $\binom{1}{3}$, labelling the new shape B. They translate image B $\binom{-2}{-6}$ and label this image C. *What translation takes A directly to C?* $\binom{-1}{-3}$ *How could you work this out from the two original column vectors?*

- Display Example 7 using ActiveTeach, showing a reflection followed by a rotation. *Which other transformation combinations could you do?*

- Ask one group of students to draw a flag in the first quadrant, a second group to draw the same flag in the in the second quadrant. Then ask them to reflect their flag in the y-axis and then the x-axis. *What single transformation is equivalent to this double reflection?* (Rotation 180° about the origin.)

Common mistakes and misconceptions

Students don't know where to start for 'general' questions such as Exercise 28G Q4–6. Encourage them to start by carrying out the transformations on different shapes, to see what happens.

Plenary

Work through Q9 as a class. What is the rule for two reflections in vertical parallel lines in the first quadrant? Does this work for reflections in horizontal parallel lines in the first quadrant? Try objects between the two mirror lines, as well as outside the mirror lines.

Modular and linear specification reference

G1.7

Keywords

transformation

Resources

tracing paper, coordinate grids to save time copying

Guided Practice Worksheet

ActiveTeach Resources

Grade Studio: Knowledge check

Links

Follow up

Middle Practice Book 28.4

Objectives
- Calculate the circumference of a circle **D**
- Calculate the perimeter of compound shapes involving circles or parts of circles **C**

Prior knowledge

Students should be able to round to a given number of significant figures or decimal places, and be able to solve one-step algebraic equations.

Starter

Write this number on the board: 3.141 592 654. Ask students to round it to:

a 1 d.p. (3.1)   **b** 2 d.p. (3.14)   **c** 3 d.p. (3.142)

d 1 s.f. (3)   **e** 2 s.f. (3.1)   **f** 3 s.f. (3.14)

Write this amount on the board: 3456.45 cm. Ask students to round it to:

g the nearest km (0 km)   **h** the nearest m (35 m)

i the nearest cm (3456 cm)   **j** the nearest mm (34 565 mm)

Main teaching

- Ask students to draw a circle of any size on a sheet of A4 paper.
  - Ask them to measure the diameter of the circle and write it down.
  - Show them how to measure the circumference with a piece of string.
  - Record the results in a table on the board showing diameter and circumference. Discuss the relationship between the two.
  - Introduce the formula for finding the circumference of a circle.
- Display Examples 1 and 2 using ActiveTeach and work through.
  - Explain that sometimes we need to calculate the radius or diameter given the circumference.
- Display Example 3 and work through.
- Display Example 4 and discuss compound shapes.

Common mistakes and misconceptions

Students forget to multiply by 2 when the radius is given and they need to use diameter.

Plenary

A bicycle has a wheel with radius 30 cm.

- *How far does the bicycle travel in one revolution of the wheel?* (188.5 cm)
- *If the wheels turn through 150 revolutions how many metres has the bicycle travelled?* (282.7 m)
- *How many revolutions will be needed for the bicycle to travel 1 km?* (531 revs)

Modular and linear specification reference
G1.5, G4.1, G4.1h, G4.3

Keywords
circumference, diameter, radius

Resources
mini-whiteboards, string

Guided Practice Worksheet
29.1 Circumference of a circle

ActiveTeach Resources
BBC Active video clips (×7)
Topic Tutor

Links
none

Follow up
Middle Practice Book 29.1

<table>
<tr><td>

Objectives
- Calculate the area of a circle **D**
- Calculate the area of compound shapes involving circles or parts of circles **C**

</td></tr>
</table>

Prior knowledge

Students should be able to round to a given number of significant figures or decimal places, and be able to substitute values into a formula and follow the order of operations

Starter

On the board display the shapes and formulae below and ask students to match the shape to the correct formula for finding the area.

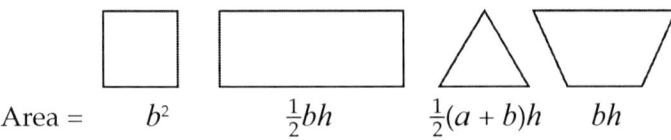

Area = b^2 $\frac{1}{2}bh$ $\frac{1}{2}(a + b)h$ bh

Main teaching

- Remind students of how to find the circumference of a circle.
  - Display a whole circle. *If the radius of the circle is r what is the circumference?* ($2\pi r$)
  - Show students how to calculate the area of the circle by making a parallelogram from sectors of the circle and then calculating the area of the parallelogram, to get the formula: area = πr^2.
- Display Examples 5 and 6 using ActiveTeach.
- Display Example 7 and discuss how sometimes it is necessary to consider parts of a circle or more than one circle.

Common mistakes and misconceptions

Students will often multiply by π before squaring.

Plenary

Check and discuss students' answers to Exercise 29E Q5. Ask them to predict the area of a further ring (of width 2 cm) drawn on the outside of the shape. (36π cm^2)

Change all the measurements to 3 cm and ask students what they notice. (Each concentric ring is 18π cm^2 larger than the last.)

Modular and linear specification reference

G1.5, G4.1, G4.3

Keywords
none

Resources
mini-whiteboards

Guided Practice Worksheet
none

ActiveTeach Resources
Animations (×2)

Links
none

Follow up
Middle Practice Book 29.2

Modular and linear specification reference

G2.4, G4.1, G4.3, G4.4

Keywords

cylinder, total surface area, curved surface area

Resources

kitchen roll (for Main teaching)
soft drink can (for Plenary)

Guided Practice Worksheet

29.3 Cylinders

ActiveTeach Resources

BBC Active video clip
Grade Studio: Knowledge check
Grade Studio: Problem solving
Topic Tutor

Links

Follow up

Middle Practice Book 29.3

Objectives

- Calculate the volume of a cylinder **C**
- Solve problems involving the surface area of cylinders **C**

Prior knowledge

Students should be able to calculate the area and circumference of a circle, and know how to draw the net of a cylinder.

Starter

Work out the circumference and area of a circle in terms of π when:

 a radius = 3 ($C = 6\pi$; $A = 9\pi$)   **b** radius = 5 (10π; 25π)
 c radius = 10 (20π; 100π)   **d** radius = 1 (2π; π)
 e diameter = 12 (12π; 36π)   **f** diameter = 24 (24π; 144π)

What size circle has circumference equal to area? (radius = 2)

Main teaching

- Remind students that a cylinder is a prism, and elicit the fact that volume of a cylinder = cross section × length = $\pi r^2 h$.

  o Display a cylinder and as a class work out the volume. Review conversions between cm^3 to ml.

- Display Examples 8, 9 and 10 using ActiveTeach.

- Display the kitchen roll and explain that they are going to find the surface area of the curved surface. Establish what the surface area is, and cut the tube and unroll to show the rectangle.

  o Establish that the width of the rectangle is equal to the circumference of the circle giving: width = $2\pi r$, and so the area of the curved surface = $2\pi rh$.

- Discuss how you might find the surface area of a closed cylinder, including the areas of both the top and bottom.

  o Establish that the surface area of the closed cylinder = $2\pi rh + 2\pi r^2$

- Display Example 11 and work through.

Common mistakes and misconceptions

Students will often multiply by π before squaring.

Plenary

Display a drink can containing 330 ml of fluid. Explain that the height of the can is 11.5 cm. *Work out the radius of the can.*

Discuss with students whether it is sensible for the can to have exactly this radius. (There should be some extra space to prevent spillage when the can is opened.)

30.1 / Converting areas and volumes

Objectives
- Convert between different units of area **D**
- Convert between different units of volume **C**

Prior knowledge

Students should be able to multiply and divide by powers of 10 confidently.

Starter

Play 'Quick on the draw'. *Draw a rectangle with an area of 60 cm².* Students must draw a rectangle with the given area using a ruler and paper. When they have finished they raise their hands. The first four to raise their hands go into the playoff: their rectangles are measured and the most accurate is the winner.

Main teaching

- Ask students to measure one large square on their graph paper.
  - Observe that one large square has an area of 1 cm².
  - Observe that one small square has an area of 1 mm².
  - Ask students to count the number of small squares in one large square. (100) Explain that 1 cm² = 100 mm².
  - Draw a square on the board and label the sides 1 m. *What is the length of each side in cm? (100 cm) What is the area of the square in cm²?* (10 000 cm²) Explain that 1 m² = 10 000 cm².
- Display Example 1 using ActiveTeach and work through.
- Sketch a cube on the board and label the sides 1 m.
  - *What is the volume of this cube? (1 m³) How many cm³ cubes would I need to use to construct this cube? (1 000 000)* Explain that 1 m³ = 1 000 000 cm³.
  - Use wooden hundred squares to illustrate that a 10 × 10 × 10 cube contains 1000 small cubes. Explain that 1 cm³ = 1000 mm³.
- Display Example 2 and work through.

Common mistakes and misconceptions

Students often convert from m³ to cm³ by multiplying by 100. Encourage students to visualise a 1 m³ block and a 1 cm³ block.

Notes on some problem-solving questions

When calculating areas and volumes of given shapes offer students the alternative strategy of converting dimensions first. Students should compare answers obtained using different strategies.

Plenary

Illustrate that 1 m³ is a large volume by asking pupils to estimate volumes of items such as backpacks, pencil cases, rubbish bins in m³. Observe that 1 m³ of water weighs 1 tonne.

Modular and linear specification reference
G3.4, G3.7

Keywords
convert, area, volume

Resources
blank paper and rulers, graph paper, wooden hundred squares (optional)

Guided Practice Worksheet
30.1 Converting areas and volumes

ActiveTeach Resources
Topic Tutor

Links
none

Follow up
Middle Practice Book 30.1

Modular and linear specification reference

G3.7

Objectives
- Calculate average speeds **D**

Prior knowledge

Students should be able to work with time confidently, and convert between times given in hours, minutes and seconds.

Starter

Practise changing the subject of a formula. Display three formulae on the board and ask students to identify the odd one out by rearranging:

$$P = Q + 2R \qquad Q = \frac{P}{2} + R \qquad R = \frac{P - Q}{2}$$

Main teaching

- *What unit can speed be measured in?* (m/s, km/h, mph)
  - Explain that the units give you a clue about how to calculate speed. *Speed is a measurement of the distance travelled in a given time.*
  - Display the formula: $\text{speed} = \frac{\text{distance}}{\text{time}}$
  - Ask students to rearrange the formula to make distance the subject.
  - Ask students to rearrange the formula to make time the subject.
  - Explain that all three formulae can be summarised using the formula triangle for distance, speed and time.
- Display Example 3 using ActiveTeach.
  - Explain that the units given in the question will determine the units of speed in the answer. *If you need to calculate the speed in different units you should convert the values before you do your calculation.*
  - Explain that questions will always say average speed or constant speed, because the calculation doesn't take into account any changes in speed over the 400 m distance.

Common mistakes and misconceptions

Students will often struggle to remember the formulae. Encourage them to write the formula triangle in their working for each question.

Plenary

An aeroplane travels from London to New York. The distance is 3430 miles. The journey takes 7 hours. What is the average speed of the aeroplane in mph? In km/h?

Keywords

speed, average speed, distance, time

Resources

Guided Practice Worksheet

30.2 Speed

30.3 Density

(combined)

ActiveTeach Resources

BBC Active video clips (×4)

Grade Studio: Problem solving

Topic Tutor

Links

http://www.sycd.co.uk/dtg/

A related resource using video of the motion of footballers to draw distance–time graphs.

Follow up

Middle Practice Book 30.2

Modular and linear specification reference

G3.7

Keywords

density, volume, mass

Resources

two (or more) identical bottles containing substances of different densities, e.g. water, sand, packing foam, 1 litre bottle of water, scales

Guided Practice Worksheet

30.2 Speed

30.3 Density (combined)

ActiveTeach Resources

BBC Active video clip

Topic Tutor

Links

Follow up

Middle Practice Book 30.3

Objectives

- Make calculations using density **D**

Prior knowledge

Students should be able to divide using a calculator and round their answers to a given degree of accuracy. They should also be able to substitute into a formula confidently.

Starter

Practise converting units of mass. Display a mass, such as 4.8 kg. Ask students to convert this mass into grams and tonnes.

Main teaching

- Display identical bottles containing substances of different densities.

  o *Do these bottles have the same volume? Do you think they have the same mass?* Invite students to weigh the bottles in turn.

  o Explain that the heavier bottle(s) are filled with material with a greater density. *Density is a measure of how much mass is contained in a given volume.*

  o Display the formula: $\text{density} = \frac{\text{mass}}{\text{volume}}$

  o Ask students to rearrange the formula to make mass the subject, and then volume the subject.

  o Explain that all three formulae can be summarised using the formula triangle for density, mass and volume.

- Calculate the density of water.

  o Use a bottle of water of known volume. *How do you convert litres into cm^3?* (×1000) Demonstrate the use of the formula to calculate the density of water.

- Display Example 4 using ActiveTeach.

  o Explain that the units given in the question will determine the units of density in the answer.

Common mistakes and misconceptions

Students will often struggle to remember the formulae. Encourage students to write the formula triangle in their working for each question.

Plenary

Explain that any object will float in a liquid with greater density. *The density of liquid mercury is 13.5 g/cm^3. A pound coin has a mass of 9.5 g and a volume of 1.25 cm^3. Will it float in mercury?*

30.4 Dimension theory

Modular and linear specification reference

G3.7

Keywords

length, area, volume

Resources

Guided Practice Worksheet

ActiveTeach Resources

BBC Active video clip

Grade Studio: Knowledge check

Topic Tutor

Links

Follow up

Middle Practice Book 30.4

Objectives

- Recognise formulae for length, area or volume by considering dimensions **B**

Prior knowledge

Students should be able to work with algebraic expressions confidently.

Starter

Ask students to call out names of 2-D or 3-D shapes beginning with different letters of the alphabet.

Main teaching

- Ask students to call out different units of length, area and volume. List them in a table with columns for length, area and volume.

  o *What do you notice about the units in each column?*

  o Ask students to recall formulae for perimeters, areas and volumes of familiar 2-D and 3-D shapes. Write these formulae in the appropriate columns.

  o Explain that lines and lengths are 1-dimensional, surfaces and areas are 2-dimensional and volumes are 3-dimensional.

  o Explain that constants have no dimensions. *You can think of a constant as a point on a number line.*

  o Use a diagram like this to illustrate 0-, 1-, 2- and 3-dimensionality.

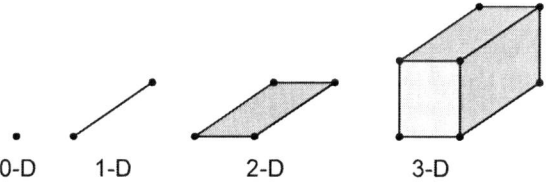

0-D   1-D   2-D   3-D

- Display Example 5 using ActiveTeach

  o Explain that length + length = length, but length × length = area.

  o Explain that dimensions can tell you whether a formula is definitely wrong, but not if it is definitely right.

Common mistakes and misconceptions

Students should take great care when some variables represent lengths, some areas and some volumes. Students can simplify their analysis by removing constants and substituting in the correct units for each variable.

Plenary

Ask students to use the SI units of time (s), length (m) and gravitational acceleration (ms^{-2}) to show that the equation for the period of a pendulum,

$$T = 2\pi \sqrt{\frac{l}{g}}$$

is consistent with dimension theory.

31.1 / Enlargement

Modular and linear specification reference

G1.7, G3.2

Objectives

- Enlarge a shape on a grid **E**
- Enlarge a shape using a centre of enlargement **D**

Prior knowledge

Students should be able to find perimeters of simple plane shapes, and be able to plot coordinates all four quadrants.

Starter

Have a few simple plane shapes on a grid on the whiteboard and ask students to find their perimeters.

Main teaching

- Display Example 1 using ActiveTeach and explain how the enlargement is drawn by multiplying each length by the scale factor.
  - After the students have completed Exercise 31A check that they understand the effect of an enlargement on perimeter.
- Display Example 2 and show how the enlargement is constructed.
  - Explain that if we are working on a grid we can locate the corners of the enlarged shape by counting squares, rather than measuring along a slanting line.
- Display Example 3. Explain that a centre of enlargement can be inside the shape or on an edge. Stress that the rules for finding the enlargement are just the same.

Common mistakes and misconceptions

The most common problem is that students are not sufficiently careful, either when copying a diagram from the book or when counting squares.

Plenary

Invite a student to do an enlargement of a shape on the whiteboard. Put a shape and its enlargement on the whiteboard and ask where the centre is and what the scale factor is. Put a shape and a reflection of its enlargement (i.e. make an 'impossible' enlargement) and ask students where the centre is – leading to a discussion of how to recognise enlargements that are possible and those that are not.

Keywords

enlargement, scale factor, multiplier, proportion, similar, centre of enlargement

Resources

plane shapes drawn on grid on whiteboard (for Starter)

squared paper, pencils, rulers

Guided Practice Worksheet

31.1 Enlargement

ActiveTeach Resources

BBC Active video clip

Topic Tutors (×2)

Links

http://www.rfbarrow.btinternet.co.uk/htmks3/Enlarge1.htm

A good demonstration of enlargements with a centre.

Follow up

Middle Practice Book 31.1

Modular and linear specification reference

G1.7h, G3.2

Keywords

Resources

squared paper, pencils, rulers

Guided Practice Worksheet

31.2 Enlargements with fractional scale factors

ActiveTeach Resources

Animation

Links

http://www.rfbarrow.btinternet.co.uk/htmks3/Enlarge1.htm

A good demonstration of enlargements with a centre.

Follow up

Middle Practice Book 31.2

Objectives

- Enlarge a shape using a fractional scale factor **C**

Prior knowledge

Students should be able to multiply a whole number by a fraction.

Starter

Invite the class to do a few multiplications of a whole number by a fraction.

Main teaching

- Display Example 4 using ActiveTeach.
  - The example shows an enlargement with scale factor $\frac{1}{3}$. The students should understand that $\times \frac{1}{3}$ is the same as $\div 3$, so the image is one third the size of the object.
  - Check they understand what an enlargement with scale factor 1 would do.

Common mistakes and misconceptions

The most common problem is that students are not sufficiently careful, either when copying a diagram from the book or when counting squares.

Plenary

Put up on the whiteboard an enlargement question with a negative fractional scale factor and invite the class to tell you the coordinates of the corners of the image without drawing the rays, forcing them to count squares.

Modular and linear specification reference
G1.7h, G1.8

Objectives
- Understand similarity and the link with enlargement **B**

Keywords
similar, ratio

Resources
pencils, rulers, calculators

Guided Practice Worksheet
31.3 Similarity

ActiveTeach Resources
Grade Studio: Knowledge check
Grade Studio: Problem solving
Topic Tutors (×2)

Links
none

Follow up
Middle Practice Book 31.3

Prior knowledge

Students should have a good understanding of ratios and be able to write ratios in their simplest form.

Starter

Put some ratios on the whiteboard and invite the class to simplify them.

Main teaching

- Display Example 5 using ActiveTeach.
  - Remind the class that enlargements produce similar figures, so the ratio of any pair of corresponding measurements will be the same, and will be equal to the enlargement scale factor.
  - Work through the example showing how we first find the ratio of lengths in its simplest form. We can use this to find the scale factor and then we may be able to find missing lengths on the diagram.
  - Part **c** of the example reminds the students that angles are preserved in an enlargement.

Common mistakes and misconceptions

Students can make errors simplifying the ratio. They sometimes choose two lengths for the ratio that are not corresponding (this is unlikely to happen in Exercise 31D because most of the shapes have been drawn the same way round).

Plenary

Display a pair of similar triangles that are not the same way round – one could be a rotation or reflection of the other. Invite the class to find the scale factor and perhaps a missing length.

Objectives

- Draw quadratic graphs **D**, **C**
- Identify the line of symmetry of a quadratic graph **C**
- Draw and interpret graphs in real-life contexts **D**, **C**

Prior knowledge

Students should be able identify the y-intercepts of straight-line graphs, recognise linear functions of the form $x = a$ and $y = a$; and substitute positive and negative numbers into expressions involving squares.

Starter

Test students' knowledge of squares of numbers from 0 to 10. Extend to squaring negative numbers from -1 to -10.

Main teaching

- Display Example 1 using ActiveTeach.
  - Emphasise that the procedure for drawing a quadratic graph is similar to that for drawing the graphs of linear functions. Summarise the procedure as:
    - Complete the table of values.
    - Plot the points.
    - Join the points with a smooth curve.

- Display Example 2.
  - Explain to students that the table of values in Example 2 have been broken down to help them work out the terms one at a time. Emphasise that in an exam, only tables of x- and y-values are given (see the exam-style questions in the Review exercise). Stress also that they will only be required to work out some of the y-values.

- Display Example 3.
  - Show how the line of symmetry passes through the lowest point of the curve. *Can you identify the line of symmetry from the table of values?*

- Display Example 4.
  - Discuss how quadratic graphs can be used to represent real-life situations.

Common mistakes and misconceptions

When a graph has its vertex between two plotted points, students often draw the bottom of the graph flat. Stress that they are aiming for a smooth curve. This may need practice.

Plenary

Display the graphs of $y = x^2$, $y = -x^2$, $y = x^2 + 1$ and $y = x^2 - 1$. Ask students to identify the maximum/minimum value of y, the x-intercept and the line of symmetry.

Modular and linear specification reference

N6.11h, N6.13

Keywords

quadratic function, curve, parabola, intercept, maximum value, minimum value

Resources

prepared graphs of $y = x^2$, $y = -x^2$, $y = x^2 + 1$, $y = x^2 - 1$ (for Plenary)

Guided Practice Worksheet

32.1 Graphs of quadratic functions

ActiveTeach Resources

BBC Active video clips (×3)
Topic Tutors (×2)

Links

Follow up

Middle Practice Book 32.1

Objectives

- Use a graph to solve quadratic equations **C**

Prior knowledge

Students should be able draw the graphs of linear functions, and find the coordinates of the point of intersection of two linear graphs.

Starter

Ask students to solve the equation $x^2 - 4 = 0$. Then ask them to draw the graph of $y = x^2 - 4$.

Main teaching

- Display the graph of $y = x^2 - 4$ from the Starter activity.
  - ○ *How could you use the graph to solve the equation $x^2 - 4 = 0$?* Explain that they are trying to solve the equation $y = 0$ and that $y = 0$ is the x-axis.
  - ○ Explain that there are usually two solutions to the equation, given by the x-coordinates of the points where the graph crosses the x-axis. Emphasise, however, that some quadratic graphs just touch the x-axis once (one solution), and some do not cross the x-axis at all (no solution). (See Common mistakes and misconceptions.)
- Display Example 5 using ActiveTeach.
  - ○ Explain that solving $x^2 - 4x = 0$ means finding where the curve crosses the x-axis ($y = 0$). Solving $x^2 - 4x = 2$ means finding where the curve crosses the line $y = 2$.
  - ○ *How would you solve $x^2 - 4x = 4$?*
- Display Example 6.
  - ○ *How would you solve $x^2 + 2x - 5 = 2$?*

Common mistakes and misconceptions

Focus should be on drawing a smooth curve. Continued practice is essential.

Students often forget to write down all the solutions. Encourage them to highlight all points of intersection on the graph.

Plenary

Display some graphs of quadratic functions and ask students to use them to solve a range of quadratic equations.

Modular and linear specification reference

N6.7h

Keywords

quadratic equation

Resources

prepared graphs of quadratic functions (for Plenary)

Guided Practice Worksheet

ActiveTeach Resources

Animation

Grade Studio: Problem solving

Topic Tutor

Links

Follow up

Middle Practice Book 32.2

Modular and linear specification reference

N6.8h

Keywords

cubic function

Resources

Guided Practice Worksheet

ActiveTeach Resources

Grade Studio: Knowledge check

Topic Tutors (×2)

Links

Follow up

Middle Practice Book 32.3

Objectives

- Draw cubic graphs **B**
- Use a graph to solve cubic equations **B**

Prior knowledge

Students should be able to substitute positive and negative numbers into expressions involving cubes and squares.

Starter

Display a partly completed table of values for $y = x^2 - 5x + 2$. Ask students to complete the table and draw the graph.

Main teaching

- Introduce students to the cubic function $y = x^3$ through consideration of its graph.
- Display Example 7 using ActiveTeach.
  - Reiterate the graph-drawing procedure:
    Complete the table of values.
    Plot the points.
    Join the points with a smooth curve.
  - It may be worth highlighting that quadratic graphs, highest power x^2, bend once, and cubic graphs, highest power x^3, bend twice.
  - Explain that for cubic functions, completion of the table of values may require more calculator use. Ensure that students are able to use the x^3 function on their calculators.

Common mistakes and misconceptions

Finding the square and cube of a negative number can cause difficulties. Highlight the difference between -1^2 and $(-1)^2$. *Is there a difference between -1^3 and $(-1)^3$?*

As with the graphical solution of quadratic equations, students often forget to write down all the solutions. Encourage them to highlight all points of intersection on the graph, and bear in mind the maximum number of possible solutions.

Plenary

It is essential that students are able to recognise the basic shape of the graphs of linear, quadratic and cubic functions, and understand how the y-intercept is reflected in the function.

Ask students to sketch a range of graphs:

    a $y = x$ **b** $y = -x$ **c** $y = x^2$ **d** $y = -x^2$ **e** $y = x^3$

What would happen if you put '+ 2' on the end of each function?

What would happen if you put '−2' on the end of each function?

Modular and linear specification reference

G3.8, G3.10

Keywords

construction, arc, perpendicular, line segment, perpendicular bisector, bisect, angle bisector

Resources

rulers, compasses, protractors (for checking only)

Guided Practice Worksheet

33.1 Constructions

ActiveTeach Resources

Animation

Topic Tutors (×2)

Links

http://www.cut-the-knot.org/ Outline/Geometry/PerpBisector. shtml

This website discusses different methods of constructing a perpendicular bisector.

http://nrich.maths.org/ public/leg.php?group_ id=39&code=168#results

This page from nrich maths has several activities related to constructions.

Follow up

Middle Practice Book 33.1

Objectives

- Construct perpendiculars **C**
- Construct the perpendicular bisector of a line segment **C**
- Construct angles of 90° and 60° **C**
- Construct the bisector of an angle **C**

Prior knowledge

Students should know how to construct triangles and estimate mid-points of line segments.

Starter

Using your compasses, draw a circle. Without changing the radius, draw a second circle, with centre on the circumference of the first, and then a third, with centre on one of the intersection points. Keep repeating to create a symmetrical pattern. What do you notice?

This Starter provides good practice in keeping the radius of the compasses constant, which is essential for the constructions that follow.

Main teaching

- Display Example 1 using ActiveTeach. *Why is it important that you do not change the radius of your compasses in Steps 1 and 2 between drawing the two arcs from P, and from A and B? What would be the result if you did?*

- Display Example 2 and work through, noting similarities with Example 1.

- Display Example 3 and work through, noting similarities and differences with Example 1. The construction of a perpendicular bisector of *AB* (Example 3) is similar to the construction in Example 1. In Example 3, the points *A* and *B* are given and the bisector must go straight through the line *AB*, so arcs must be constructed both above and below the line.

- Display Example 4. Ask one of your more confident students to explain the method to the rest of the class.

- Display Example 5. Explain that an angle bisector should be a continuous line that goes in both directions from the given angle, rather than just starting at the angle and going in the direction of the two lines.

Common mistakes and misconceptions

Students may fail to keep settings of compasses constant.

Plenary

Sketch this triangle on the board.

Is it possible to construct this triangle?

If not, explain why.

(Hint: Construct a square on the base line, with diagonal of 10 cm. The sides of the square must be slightly over 7 cm ($\sqrt{50}$ m). So a side as shown of exactly 7 cm would not meet the base line.)

Objectives

- Construct loci **C**
- Solve locus problems, including the use of bearings **C**

Prior knowledge

Students should be able to use three-figure bearings, and to visualise the paths taken by everyday moving objects.

Starter

On a frozen lake, skater A is due west of skater B. Skater A sets off on a bearing of 050°, and B sets off on a bearing of 290°. Will the skaters' paths cross? Ask students to draw a sketch to find the solution.

Main teaching

- Explain that a locus is a set of points that follow a given rule or rules, and that we can use these rules to construct the locus. The simplest loci are:
  - a circle (a set of points equidistant from a given point)
  - two parallel lines equidistant from a given line.
- Display Example 6 using ActiveTeach and work through.
- Display Example 7 to show how these loci are combined to give the racetrack shape.
- Ask the class to visualise and describe a locus equidistant from two fixed points, and from two fixed lines, before showing and explaining the explanatory text on special loci.
- Display Examples 8 and 9 and work through. Emphasise the importance of accurate constructions. In Example 9, point out that labelling points on the diagram (e.g. *X* and *Y*) help in describing the exact locus.

Common mistakes and misconceptions

Students may make mistakes with inaccurate constructions and by shading the wrong region. Encourage students to re-read the question and check their drawings.

Plenary

Jon is lost. He comes across a signpost, saying Hightown 5 miles; Lowboro 3 miles. Can he pinpoint on his map where he is? Does he need more information? If so, what? (A bearing, or a third place on the signpost.)

Draw a diagram to illustrate your answer. Next, use an atlas to choose a position, and give your partner some information to help them find the exact position.

Modular and linear specification reference

G3.11

Keywords

locus (loci), equidistant

Resources

ruler, compasses, protractors
atlases (for Plenary, optional)

Guided Practice Worksheet

33.2 Locus

ActiveTeach Resources

BBC Active video clip
Grade Studio: Knowledge check
Grade Studio: Problem solving
Topic Tutors (×2)

Links

http://nrich.maths.org/public/leg.php?group_id=39&code=-330#results
This site has many Resources related to loci, some of which are interactive.

Follow up

Middle Practice Book 33.2

34.1 Pythagoras' theorem

Objectives
- Understand Pythagoras' theorem **C**

Prior knowledge

Students should know how to find the squares and square roots of numbers with and without a calculator.

Starter

Ask quick-fire questions starting off with finding the squares of 5, 7, 2, etc., and move on to decimals and larger numbers with and without a calculator. Move on to finding the square roots of numbers with or without a calculator. (Hand-held mini-whiteboards are useful for this activity.)

Main teaching

- Display the explanatory text showing the triangle with the squares drawn on each side using ActiveTeach.
  - Ask students to work out the area of each square in a 3, 4, 5 triangle. Discuss what they notice.
  - Explain the connection between the sides of a right-angled triangle and the definition of Pythagoras' theorem.
- Display Example 1 using ActiveTeach and work through.
  - Ensure that all students are clear which side of the triangle is the hypotenuse.

Common mistakes and misconceptions

Students forget that x^2 means $x \times x$, not $x \times 2$. Practise with numbers other than 2.

Plenary

Go through answers to both questions in Exercise 34A. Ask students to make up a question similar to those in the exercise, work out the answer, and then swap their question with another student.

Choose two students to share their questions and answers with the rest of the class.

Modular and linear specification reference

G 2.1

Keywords

Pythagoras' theorem, right-angled triangle, hypotenuse

Resources

mini-whiteboards (for Starter)

Guided Practice Worksheet

ActiveTeach Resources

Animation

BBC Active video clip

Links

Follow up

Middle Practice Book 34.1

Objectives

- Calculate the hypotenuse of a right-angled triangle **C**
- Solve problems using Pythagoras' theorem **C**

Modular and linear specification reference

G2.1

Keywords

Resources

Guided Practice Worksheet

ActiveTeach Resources

Topic Tutor

Links

Follow up

Middle Practice Book 34.2

Prior knowledge

Students need to know how to find the squares and square roots of numbers with and without a calculator, and should be able to substitute values into a formula. They also need to know which side of a right-angled triangle is the hypotenuse.

Starter

Draw some right-angled and non-right-angled triangles on the board. Ask students to label the hypotenuse. (Draw the triangles in a variety of orientations.)

Main teaching

- Display Example 2 using ActiveTeach. Ask students to work through the example in pairs. If they are finding it difficult, do an example first where the length of the hypotenuse is a whole number, such as a 3, 4, 5 triangle. Recap on rounding to a suitable degree of accuracy, usually one decimal place or two significant figures.

- Display Example 3. Ask pairs of students to try and solve the problem.
  - Discuss how to solve and break down a problem-type question.

Common mistakes and misconceptions

The most common mistake is forgetting to take the square root to find the final answer. Emphasise that if an exam question says to calculate a length of a hypotenuse then a scale diagram would receive no credit.

Plenary

Go through the answers to both Exercise 34B and 34C.

Choose a more complex question, such as Exercise 34C Q7 or Q8. Ask students to solve this problem (they could work in pairs or groups). Ask one group to share their solution with the class. Emphasise that all steps in their working should be shown.

<table>
<tr><td>

Objectives
- Calculate the length of a shorter side in a right-angled triangle **C**
- Solve problems using Pythagoras' theorem **C**

</td></tr>
</table>

Prior knowledge

Students need to know how to find the squares and square roots of numbers with and without a calculator, and should be able to substitute values into a formula. They also need to know which side of a right-angled triangle is the hypotenuse.

Starter

Display a set of whole numbers and ask students to identify the square numbers in it. Try to get the first 12 square numbers and 625, 169 and 225. Ask for the squares of some simple decimals, such as 0.5, 0.1, etc.

Main teaching

- Display Example 4 using ActiveTeach. This shows the lengths of the hypotenuse and one of the shorter sides. Ask students to work in pairs and find the length of the third side of the triangle.
  - Take feedback and emphasise that it is easier to write out the theorem and substitute in values than rearrange the formula first.
  - Display a second example and ask students to work through it. Take feedback and check that sufficient working is being shown.
- Display Example 5 and ask students to work out the perpendicular height.
  - Emphasise that Pythagoras' theorem only works with right-angled triangles. *How can we make this a right-angled triangle*? Discuss how to solve this question.

Common mistakes and misconceptions

The most common mistake is forgetting to take the square root to find the final answer. Emphasise that Pythagoras' theorem only applies to right-angled triangles. Students must draw a diagram carefully and make sure it is right-angled to solve the problem.

Plenary

Choose a more complex question such as Exercise 34E Q6 or Q7. Ask students to solve this problem (in pairs or groups). Ask one group to share their solution with the class.

Modular and linear specification reference

G2.1

Keywords

Resources

Guided Practice Worksheet

ActiveTeach Resources

Animation

Grade Studio: Problem solving

Topic Tutor

Links

Follow up

Middle Practice Book 34.3

Objectives

- Calculate the length of a line segment AB **C**

Modular and linear specification reference

G2.1

Prior knowledge

Students should be able to use Pythagoras' theorem to find the length of a hypotenuse.

Keywords

line segment

Resources

Guided Practice Worksheet

ActiveTeach Resources

Grade Studio: Knowledge check

Links

Follow up

Middle Practice Book 34.4

Starter

Display a coordinate grid which goes from –8 to 8 on both axes. Plot 10 random points labelled *A–L* and ask students to write down the coordinates.

Main teaching

- Display Example 6 using ActiveTeach.

  o *How can we work out the length of the line segment?* Take feedback and discuss the ways in which students found the length.

  o Emphasise that a diagram is useful when helping to visualise when the two sets of coordinates are used as the end-points. Draw in a right-angled triangle from these points.

- Display a second example where only the coordinates are given. Ask students to work in pairs to find the length of the line segment.

Common mistakes and misconceptions

A common mistake is to add instead of subtract the two pairs of coordinates.

Plenary

On the board, display a question using only negative coordinates. Ask students to solve this problem and represent their solution with a diagram (they could work in pairs or groups). Ask one group to share their solution with the class.

Objectives
- Understand and recall trigonometric ratios in right-angled triangles **B**
- Know how to enter the trigonometric functions on a calculator **B**

Modular and linear specification reference

G2.2h, N1.14h

Keywords
trigonometry, sine, cosine, tangent, hypotenuse, opposite, adjacent, inverse

Resources
none

Guided Practice Worksheet
none

ActiveTeach Resources
none

Links
none

Follow up
Middle Practice Book 35.1

Prior knowledge

Students need to know which side of a right-angled triangle is the hypotenuse, and be able to substitute values into a formula.

Starter

Draw two right-angled triangles on the board. Ask students to label the hypotenuse. *What are the two key points about a hypotenuse?* (It is opposite the right angle; it is the longest side.)

Main teaching

- On each of the two triangles used in the Starter, add in an angle labelled x in different positions. Explain that trigonometry is used for calculating missing lengths and angles in right-angled triangles.

  o On the first triangle show how to label the sides opposite, adjacent and hypotenuse. Ask students to do the same for the second triangle.

  o Explain that the ratios of the sides can be represented by sine, cosine and tangent. Go through the ratios and the mnemonic SOHCAHTOA.

- Display the explanatory text on the use of calculators and work through as a class. This is often quite tricky because of the variety of different calculators.

- Display Example 1 using ActiveTeach and ensure that all students know how to use their calculators correctly for this application.

Common mistakes and misconceptions

Emphasise that the sine, cosine and tangent ratios only apply to right-angled triangles. Students need to draw a diagram carefully and make sure the triangle is right-angled to solve the problem.

Plenary

Go through answers to Exercise 35A. Make sure that all students can do these.

Ask each student to write down five questions similar to those in the exercise. They exchange their questions, work out the answers to their partner's questions, swap again, and check/mark the answers.

Modular and linear specification reference
G2.2h

Keywords
none

Resources
none

Guided Practice Worksheet
35.2 Finding lengths using trigonometry

ActiveTeach Resources
Animation

Links
none

Follow up
Middle Practice Book 35.2

Objectives

- Use trigonometric ratios to find lengths in right-angled triangles **B**

Prior knowledge

Students need to know which side of a right-angled triangle is the hypotenuse. They should also be able to substitute values into a formula.

Starter

Write down some equations of the form $\frac{x}{3} = 5$ or $\frac{10}{x} = 2$. Ask students to solve these equations and discuss how they solved them.

Main teaching

- Display Example 2 using ActiveTeach to show students the calculations needed to work out missing lengths. Go through it step by step.
- Display three right-angled triangles with a different angle given, and a missing length to be calculated (with different trigonometric ratios).
  - Ask students to label the sides of each triangle. Look at the first triangle and decide which ratio is going to be used.
  - Ask students to work in pairs to find the missing length on the other two triangles.
- Display Example 3. Explain how to rearrange the equation to find the missing length.

Common mistakes and misconceptions

Emphasise that the sine, cosine and tangent ratios only apply to right-angled triangles. Students need to draw a diagram carefully and make sure the triangle is right-angled to solve the problem.

Plenary

Draw five triangles with a mixture of information, so students will have to use all three ratios. Have two questions with the unknown as the denominator. Write the answers on the board. Ask students to match the questions with the answers but emphasise that sufficient working has to be shown to get full marks.

Finding angles using trigonometry

Modular and linear specification reference

G2.2h

Objectives

- Use trigonometric ratios to find the angles in right-angled triangles **B**

Prior knowledge

Students need to know which side of a right-angled triangle is the hypotenuse and be able to substitute values into a formula.

Starter

Draw three right-angled triangles on the board, labelling one of the angles x in each triangle. Ask students to label the sides of the triangles with respect to x.

Main teaching

- Display Example 4 using ActiveTeach and work through, showing the calculations needed to find the required angle.
- Place some values on the three right-angled triangles used in the Starter, so that a different trigonometric ratio will needed to find the required angle in each triangle.
  - o Ask students to decide which ratio is going to be used. Take feedback.
  - o Emphasise that if all three sides are known, such as a 3, 4, 5 triangle, then any of the ratios can be used.
  - o Recap how to use a calculator to find the angle by using the inverse function. (You may choose to introduce the term $\sin^{-1}$.)
  - o Ask students to work in pairs and find the missing angles.

Common mistakes and misconceptions

Although $\sin^{-1}$ has been introduced, it is not required at GCSE. If your class is likely to become confused, it is better to leave it out and simply say that the inverse is being found.

Plenary

Draw a triangle on the board with two sides and an angle, and ask students to use trigonometry to find the third side. Check the answer using Pythagoras' theorem. Repeat for a few more triangles.

Keywords

Resources

Guided Practice Worksheet

35.3 Finding angles using trigonometry

ActiveTeach Resources

Topic Tutor

Links

Follow up

Middle Practice Book 35.3

Objectives

- Use trigonometric ratios and Pythagoras' theorem to solve problems, including the use of bearings **B**

Prior knowledge

Students should be able to use Pythagoras' theorem to calculate the length of a missing side, and should understand and be able to use bearings.

Starter

Draw a right-angled triangle with two sides given, and ask students to calculate the length of the missing side.

Draw a selection of diagrams similar to the Skills check and ask students to find the bearing of A from B and B from A.

Main teaching

- Display Example 5 using ActiveTeach. Ask students to work through this in pairs.
  - Take feedback to generate a model solution, going through it step by step.
  - Emphasise the need to draw a diagram in order to visualise the problem. Also emphasise full working should be shown.
- Depending on the need of the class, work through other examples similar to the AO2 and AO3 questions in Exercise 35D.

Common mistakes and misconceptions

Solving more problem-type questions may prove very difficult for some students. It is essential they draw a clear diagram in order to help them progress with the question.

Notes on some problem-solving questions

In Q7 and Q8, students must work through several stages before the final solution is found. Suggest to students that a plan of action is useful for these questions.

Plenary

Put Q7 or Q8 on the board and ask students to work though either question in pairs.

Either go through the solution or ask one of the pairs to share their solution. Again emphasise the need for clear working to be shown.

Modular and linear specification reference

G2.2h, G3.6

Keywords

Resources

Guided Practice Worksheet

ActiveTeach Resources

Animation

Links

Follow up

Middle Practice Book 35.4

Objectives

- Solve problems using an angle of elevation or an angle of depression **B**

Prior knowledge

Students need to be able to use trigonometry in right-angled triangles to calculate lengths and angles, and should be able to substitute values into a formula.

Starter

Draw two triangles on the board and ask students to calculate a missing length and a missing angle using trigonometry, to revise calculating lengths and angles with trigonometry. Go through and check everyone has shown full working out.

Main teaching

- Display the explanatory text and diagrams on angles of elevation and depression using ActiveTeach. Explain what is meant by the angle of elevation and the angle of depression.
  - Emphasise that the angle of depression is measured from the horizontal downwards.
- Display Example 6 and work through. Emphasise the importance of labelling the diagram correctly.

Common mistakes and misconceptions

Throughout this section, the ability to extract the information needed to draw a right-angled triangle is crucial. It may be necessary to demonstrate more examples.

Plenary

Go through a question similar to Q5. Check that all students have drawn an accurate diagram and can extract the right-angled triangle they need to solve the problem. Ask students to design their own question and answer, then swap with another student. Discuss their solutions.

Modular and linear specification reference

G2.2h

Keywords

angle of elevation, angle of depression

Resources

Guided Practice Worksheet

ActiveTeach Resources

Grade Studio: Knowledge check

Grade Studio: Problem solving

Links

Follow up

Middle Practice Book 35.5

Circle properties

Modular and linear specification reference
G1.5h

Objectives
- Use chord and tangent properties to solve problems **B**

Prior knowledge

Students should be able to use Pythagoras' theorem to find lengths of missing sides, and be able to find missing angles using simple angle rules.

Starter

Find the missing side of a triangle using Pythagoras' theorem. This could be done as a whole class activity using mini-whiteboards.

Main teaching

- Show the explanatory text on circle properties using ActiveTeach and check that students can name the parts of the circle, and understand that the perpendicular from the centre of a circle to a chord bisects the chord.
- Display Example 1 using ActiveTeach and work through. Encourage students to draw a right-angled triangle with lengths 13 cm and 12 cm to help them to solve the problem.
- Show the explanatory text on Properties 2 and 3, and discuss properties of tangents.
- Display Example 2 and explain where the tangent properties have been used.

Common mistakes and misconceptions

Students often find it difficult to begin questions that use chord and tangent properties of circle. Encourage students to clearly state any facts they do know even if they do not seem immediately relevant to help them progress with the question.

Plenary

Ask students to make a poster stating circle properties with diagrams as the lesson progresses and finish it during the plenary and for homework.

Keywords
chord, circumference, segment, arc, tangent

Resources
mini-whiteboards (for Starter)

Guided Practice Worksheet
none

ActiveTeach Resources
Animation

Links
http://www.mathwarehouse.com/geometry/circle/

Follow up
Middle Practice Book 36.1

Objectives
- Use circle theorems to solve geometrical problems **B**

Prior knowledge

Students should be able to use Pythagoras' theorem to find lengths of missing sides, and be able to find missing angles using simple angle rules.

Starter

Ask students to solve a range of problems involving angle rules, including those involving angles in quadrilaterals, triangles and around a point.

Main teaching

- Display the explanatory text for Theorem 1 using ActiveTeach.
  - *What would the angle at the circumference be if the angle at the centre was____?*
  - *What would the angle at the centre be if the angle at the circumference was ____?*
- Display Example 3 and work through it.
- Display the explanatory text for Theorem 2.
  - Explain how the second theorem is a special case of the first theorem.
- Display Example 4 and work through it.
- Display the explanatory text for Theorem 3 and work through Example 5.
- Display the explanatory text for Theorem 4.

Common mistakes and misconceptions

Students sometimes mistake chords for diameters and therefore incorrectly identify the subtended angle as 90°.

Plenary

Discuss how one theorem can often lead to another. Ask students to give examples from the lesson. Ask the more able to explain how they might use Theorem 1 to prove Theorem 2.

Modular and linear specification reference
G1.5h

Keywords
subtended, cyclic quadrilateral, supplementary

Resources
mini-whiteboards

Guided Practice Worksheet
36.2 Circle theorems

ActiveTeach Resources
Animations (×2)
Grade Studio: Knowledge check
Grade Studio: Problem solving
Topic Tutor

Links
none

Follow up
Middle Practice Book 36.2

Missed appointments

Objectives

In a practical situation be able to:

- understand dual bar charts
- calculate mean, median and mode.

Prior knowledge

Students should be able to use a calculator efficiently.

Starter

Ask students to find the mean, median and mode from a bar chart such as:

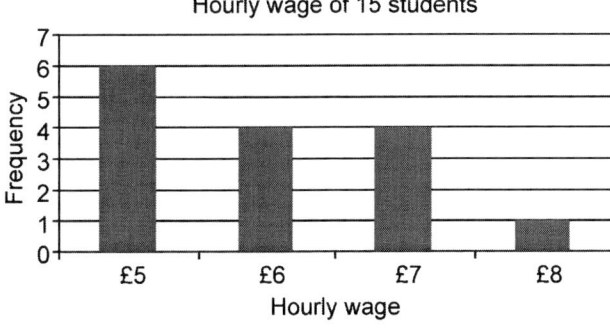

Hourly wage of 15 students

(Answers: mean = £6, median = £6, mode = £5.)

Main teaching

- Ensure students have time to read all information and questions twice.
- If students have a photocopy they should highlight the facts that may be useful. Using different colours for different questions may be useful for some students.
- If students do not have a photocopy they should be encouraged to write down all the relevant facts for each question.
- Encourage students to show full workings and give reasons and explanations whenever the need arises. Answers only are not enough.
- Rereading and checking work is an essential element of success in functional maths. Students should not be afraid of rewriting or amending parts of questions – these changes, however, must be clear.

Common mistakes and misconceptions

Not reading the information carefully and not checking answers to see if they are reasonable causes the most problems.

Students may mix up the mean, median and mode.

Students may not explain clearly enough what they have done and why. Bullet points are an acceptable method to show clarity.

Plenary

Discuss answers and where in the text they were found.

Discuss the methods used to answer the questions and also the best ways to show the examiner exactly what has been done.

Discuss key words used when trying to 'explain clearly' in Q6.

Resources

Links to Middle sets Student Book

- Understanding dual bar charts: page 46
- Calculating mean, median and mode: page 54

Functional Skills Mathematics specification references

Level 1

- Find mean and range
- Extract and interpret information from tables, diagrams, charts and graphs

Level 2

- Use statistical methods to investigate situations

Skill standards

Representing: Q1, Q3
Analysing: Q2, Q5
Interpreting: Q4, Q6

Objectives

In a practical situation be able to:

- convert between m*l* and litres
- convert between g and kg
- add time
- understand and use proportion.

Prior knowledge

Students should be able to use the unitary method for adapting recipes.

Starter

Ensure that students know that 1000 g = 1 kg and 1000 m*l* = 1 *l*.

Practise adding numbers of minutes together, such as 45 + 35 and get students to give their answers in minutes and also in hours and minutes.

Practise adapting simple recipes for different numbers of people.

Main teaching

- Make sure students have time to read all the information and questions twice.
- Students should be encouraged to show full workings and give reasons and explanations wherever the need arises. Answers only are not enough.
- Rereading and checking work is an essential element of success in functional maths. Students should not be afraid of rewriting or amending parts of questions – these changes, however, must be clear.
- Show students that Q3 is too easy if they just copy the original recipe out. The original recipe is for 10 people, they need to adapt it for 8.

Common mistakes and misconceptions

Not reading the information carefully and not checking answers to see if they are reasonable causes the most problems.

Students may not explain clearly enough what they have done and why. Bullet points are an acceptable method to show clarity.

Students may not be sufficiently thorough in Q4.

Students may not realise that one dish can be cooking while another is being prepared.

Plenary

Discuss answers and where in the text they were found. Discuss the methods used to answer the questions and also the best ways to show the examiner exactly what has been done.

Discuss any problems students had with understanding a question, and what key words were used to aid understanding.

Ask students if any of them cook, and if they do, did it help them with this exercise?

Resources

simple recipes to adapt for different numbers of people

Links to Middle sets Student Book

- Converting between m*l* and litres: page 110
- Converting between g and kg: page 110
- Understanding and using proportion: Chapter 7

Functional Skills Mathematics specification references

Level 1

- Solve problems requiring calculation with common measures, including money, time, length, weight, capacity and temperature
- Solve simple problems involving ratio

Level 2

- Use, convert and calculate using metric and, where appropriate, imperial measures
- Understand and use positive and negative numbers of any size in practical contexts

Skill standards

Representing: Q6
Analysing: Q1, Q2, Q3
Interpreting: Q4, Q5

Objectives

In a practical situation be able to:

- extract data from tables
- convert between g and kg
- solve two-step equations.

Resources

Links to Middle sets Student Book

- Extracting data from tables: Chapter 1
- Converting between g and kg: page 110
- Solving two-step equations: page 224

Functional Skills Mathematics specification references

Level 1

- Convert units of measure in the same system
- Extract and interpret information from tables, diagrams, charts and graphs

Level 2

- Understand and use simple formulae and equations involving one- or two-step operations

Skill standards

Representing: Q2, Q3
Analysing: Q1, Q5
Interpreting: Q4

Prior knowledge

Students should understand the concept of eBay.

Starter

Give students some two-step algebra equations to solve, e.g. $2x + 3 = 11$, $6x - 5 = 31$, $\frac{x}{8} + 3 = 9$.

Main teaching

- Make sure students have time to read all the information and questions twice.
- If students have a photocopy they should highlight the facts that may be useful in answering Q1–3. Using different colours for different questions may be useful for some students.
- Students should be encouraged to show full workings and give reasons and explanations whenever the need arises. Answers only are not enough.
- Rereading and checking work is an essential element of success in functional maths. Students should not be afraid of rewriting or amending parts of questions – these changes, however, must be clear.
- Q4 requires a lot of calculations. Students must be careful to make sure they are being clear and accurate with their workings, as mistakes here will be easily missed when checking.

Common mistakes and misconceptions

Not reading the information carefully and not checking answers to see if they are reasonable causes the most problems.

Having a lot of information to deal with at one time often causes issues with students who may improvise with short cuts that often do not work. Students should be encouraged to give full answers and show all workings.

Plenary

Discuss answers and where in the text they were found.

Discuss the methods used to answer the questions and also the best ways to show the examiner exactly what has been done. Look especially at students' solutions to Q4. Look for best practice within the class, highlighting work that is well set out and so is clear and easy to follow.

Objectives

In a practical situation be able to:

- read two-way tables
- read line graphs
- understand negative numbers in context
- round to the nearest 10.

Prior knowledge

Students should have a working knowledge of 12-hour and 24-hour clock times.

Starter

Discuss negative numbers. Ask questions such as: *Which temperature is warmer, –4°C or –10°C?*

Practise rounding numbers to the nearest 10, for example: 78, 515, 692, etc.

Main teaching

- Make sure students have time to read all the information and questions twice.

- If students have a photocopy they should highlight the facts that may be useful – using different colours for different questions may be useful for some students.

- If students do not have a photocopy they should be encouraged to write down all the relevant facts for each question.

- Students should be encouraged to show full workings and give reasons and explanations whenever the need arises. Answers only are not enough.

- Rereading and checking work is an essential element of success in functional maths. Students should not be afraid of rewriting or amending parts of questions. However, these changes must be clear.

Common mistakes and misconceptions

Not reading the information carefully and not checking answers to see if they are reasonable causes the most problems.

Students may not explain clearly enough what they have done and why. Bullet points are an acceptable method to show clarity.

Plenary

Discuss answers and where in the text they were found.

Discuss the methods used to answer the questions and also the best ways to show the examiner exactly what has been done.

Discuss whether it would be reasonable to buy the Warm'n'cosy sleeping bag (–5°C) even though the minimum night-time temperature **outside** the **tent** may get as low as –7°C.

Resources

Links to Middle sets Student Book

- Reading two-way tables: page 10
- Understanding negative numbers in context: page 154
- Rounding to the nearest 10: Chapter 9

Functional Skills Mathematics specification references

Level 1

- Extract and interpret information from tables, diagrams, charts and graphs

Level 2

- Understand and use positive and negative numbers of any size in practical contexts

Skill standards

Representing: Q1
Interpreting: Q2, Q3

Objectives

In a practical situation be able to:

- work out lengths from a scale drawing
- work out perimeters of squares, rectangles and circles
- work out volumes of cuboids and cylinders.

Prior knowledge

Students should be able to measure accurately and work out total costs when given price per unit, such as 15 m at £3 per metre.

Starter

Ask students to work out the perimeters of various squares, rectangles and circles. Also ask them to work out the volumes of various cuboids and cylinders.

Main teaching

- Make sure students have time to read all the information and questions twice.
- Students should be encouraged to show full workings and give reasons and explanations whenever the need arises. Answers only are not enough.
- Rereading and checking work is an essential element of success in functional maths. Students should not be afraid of rewriting or amending parts of questions – these changes, however, must be clear.

Common mistakes and misconceptions

Not reading the information carefully and not checking answers to see if they are reasonable causes the most problems.

Students may not leave a 10 cm gap at the top of the plant pots when working out the volume of compost required.

Having a lot of information to deal with at one time often causes issues with students who may improvise with short cuts that often do not work. Students should be encouraged to give full answers and show all workings.

Plenary

Discuss answers and where in the text they were found.

Discuss the methods used to answer the questions and also the best ways to show the examiner exactly what has been done. Look especially at solutions to Q5. Look for best practice within the class, highlighting work that is well set out and so is clear and easy to follow.

Resources

rulers

Links to Middle sets Student Book

- Working out lengths from a scale drawing: page 362
- Working out perimeters of squares and rectangles: page 405
- Working out the circumference of a circle: page 444
- Working out volumes of cuboids: page 411
- Working out volumes of cylinders: page 451

Functional Skills Mathematics specification references

Level 1

- Construct geometric diagrams, models and shapes

Level 2

- Find area, perimeter and volume of common shapes

Skill standards

Analysing: Q1, Q2, Q3, Q4
Interpreting: Q5

Objectives

In a practical situation be able to:

- use the distance, speed and time formula
- work out a fraction of an amount
- increase an amount by a given percentage.

Prior knowledge

Students should be able to use and understand a route on a map, and should know how to add and subtract time.

Starter

Remind students of the distance, speed and time formula:

distance = speed × time

Ask students some basic questions using the formula, such as:

- speed = 50 mph, time = 3 hours, distance = ?
- distance = 120 km, time = 4 hours, speed = ?
- distance = 100 miles, speed = 40 mph, time = ?

Main teaching

- Make sure students have time to read all the information and questions twice.
- Students should be encouraged to show full workings and give reasons and explanations whenever the need arises. Answers only are not enough.
- Rereading and checking work is an essential element of success in functional maths. Students should not be afraid of rewriting or amending parts of questions – these changes, however, must be clear.
- Some students may need prompting in Q4; the first cyclist is likely to start at 7:30 and have the fastest average speed, the last cyclist is likely to start at 8:30 and have the slowest average speed.

Common mistakes and misconceptions

Not reading the information carefully and not checking answers to see if they are reasonable causes the most problems.

Students may not explain clearly enough what they have done and why. Bullet points are an acceptable method to show clarity.

Students may not realise that the start times affect the answer in Q4.

Plenary

Discuss answers and where in the text they were found.

Discuss the methods used to answer the questions and also the best ways to show the examiner exactly what has been done. Look especially at students' solutions to Q4. Look for best practice within the class, highlighting work that is well set out and so is clear and easy to follow.

Resources

rulers

Links to Middle sets Student Book

- Working out lengths from a scale drawing: page 362
- Working out perimeters of squares and rectangles: page 405
- Working out the circumference of a circle: page 444
- Working out volumes of cuboids: page 411
- Working out volumes of cylinders: page 451

Functional Skills Mathematics specification references

Level 1

- Construct geometric diagrams, models and shapes

Level 2

- Find area, perimeter and volume of common shapes

Skill standards

Analysing: Q1, Q2, Q3, Q4

Interpreting: Q5